152
Advances in Polymer Science

Springer-Verlag Berlin Heidelberg GmbH

Viscoelasticity
Atomistic Models
Statistical Chemistry

With contributions by
J. Baschnagel, K. Binder, P. Doruker, G. Grigorescu,
A. A. Gusev, O. Hahn, K. Kremer, S. I. Kuchanov,
W.-M. Kulicke, W. L. Mattice, F. Müller-Plathe,
M. Murat, W. Paul, S. Santos, U. W. Suter, V. Tries

Springer

This series presents critical reviews of the present and future trends in polymer and biopolymer science including chemistry, physical chemistry, physics and materials science. It is addressed to all scientists at universities and in industry who wish to keep abreast of advances in the topics covered.

As a rule, contributions are specially commissioned. The editors and publishers will, however, always be pleased to receive suggestions and supplementary information. Papers are accepted for „Advances in Polymer Science" in English.

In references Advances in Polymer Science is abbreviated Adv. Polym. Sci. and is cited as a journal.

Springer APS home page: http://link.springer.de/series/aps/ or
http://link.springer-ny.com/series/aps
Springer-Verlag home page: http://www.springer.de

ISSN 0065-3195
ISBN 978-3-662-15660-5 ISBN 978-3-540-46778-6 (eBook)
DOI 10.1007/978-3-540-46778-6

Library of Congress Catalog Card Number 61642

Originally published by Springer-Verlag Berlin Heidelberg New York in 2000
Softcover reprint of the hardcover 1st edition 2000

Typesetting: Data conversion by MEDIO, Berlin
Cover: MEDIO, Berlin
Printed on acid-free paper SPIN: 10706218 02/3020hu - 5 4 3 2 1 0

Editorial Board

Advances in Polymer Science
Now Also Available Electronically

For all customers with a standing order for Advances in Polymer Science we offer the electronic form via LINK free of charge. Please contact your librarian who can receive a password for free access to the full articles. By registration at:

http://link.springer.de/series/aps/reg_form.htm

If you do not have a standing order you can nevertheless browse through the table of contents of the volumes and the abstracts of each article at:

http://link.springer.de/series/aps/

There you will find also information about the

- Editorial Bord
- Aims and Scope
- Instructions for Authors

Contents

Prediction of Viscoelastic Properties and Shear Stability of Polymers in Solution

Gabriela Grigorescu [1], Werner-Michael Kulicke [2]

[1] Institut de Genie Chimique I, Laboratoire des Polymères et Biomatériaux EPFL-Ecublens, 1015 Lausanne, Suisse
[2] Institut für Technische und Makromolekulare Chemie, Universität Hamburg, Bundesstrasse 45, D-20146 Hamburg, Germany
e-mail: kulicke@chemie.uni-hamburg.de

Dedicated to the late Professor Roger S. Porter, University of Massachusetts, USA

The rheological behaviour of dissolved polymers is very complex. Both experiment and theory in this field have undergone rapid development in recent years. Therefore, we describe the possibilities of predicting the viscoelastic properties as well as the shear stability using the entanglement and reptation concepts and exemplifying mainly with narrow distributed polystyrene samples. The viscoelastic properties are discussed in relation to molar mass, concentration, solvent quality, chemical structure and shear rate. The structure-property relationships derived here permit the prediction of both the zero-shear viscosity, η_0, as well as the shear rate dependent viscosity $\eta(\dot{\gamma})$. These relationships can be extended to non-Newtonian fluids. For solutions of coiled polymers in a thermodynamically good solvent, five distinct states of solution are formed: ideally dilute solution, semi-dilute particle solution, semi-dilute network solution, concentrated particle solution and concentrated network solution. For non-homogeneous, semi-dilute (moderately concentrated) solutions the slope in the linear region of the flow curve $[\eta=f(\dot{\gamma})]$ must be standardised against the overlap parameter $c\cdot[\eta]$. Furthermore, it is possible to predict the onset of shear degradation of polymeric liquids subjected to a laminar velocity field on the basis of molecular modeling. Also described is the phenomenon that the elastic nature (first normal stress difference) may overwhelm the viscous nature (shear stress) at relatively low shear rates. This high elasticity can cause deviation from laminar flow conditions and the onset conditions can be detected by plotting $S_R=f(\tau_{12})$.

Keywords. Viscoelasticity, Polymer solution, State of solution, Shear stability criteria, Flow irregularities, Weissenberg number

List of Symbols and Abbreviations

a	Mark–Houwink constant
$(A)_c$	constant
B_1, B_2, B_n	constants
c	concentration (g/cm^3)
c^*, c^{**}	critical concentrations (g/cm^3)
E	number of entanglements per molecule
F(r)	connector factor
F_{KPA}	normal force in raised cone-and-plate geometry (kg)
G'	storage modulus
G"	loss modulus
G_p	plateau modulus
h	shift factor (τ_0/τ_R)
K', K, K_1, K_2, K_n	constants
K_H	Huggins constant
K_η	constant of the Mark–Houwink relationship
M	molar mass
N	number of segments per molecule

N_1	first normal stress (Pa)
N_2	second normal stress (Pa)
n	slope $d(\log \eta_{sp})/d(\log [\eta])$ at high $c\cdot[\eta]$
p	p-th mode of the relaxation time spectrum
r	director vector
PAAm	polyacrylamide
PS	polystyrene
R	gas constant
R	radius of cone-and-plate geometry (m)
S_R	recoverable strain
T	temperature
β	cone angle (rad)
β	modified reduced shear rate
ϕ	volume fraction
$\dot{\gamma}$	shear rate (s^{-1})
$\dot{\gamma}R$	shear rate at the edge of raised cone-and-plate geometry (s^{-1})
η	viscosity (Pa·s)
η_0	zero-shear rate viscosity
η_s	solvent viscosity (Pa·s)
η_{sp}	specific viscosity
$\eta\dot{\gamma}$	apparent viscosity at shear rate $\dot{\gamma}$
$[\eta]$	intrinsic viscosity (cm^3/g)
ρ	density of polymer solution (kg/m^3)
τ_{12}	shear stress
ω	angular velocity ($rad\cdot s^{-1}$)
ξ	screening length
λ_0	experimentally derived relaxation time
conc	concentrated
crit	critical
deg	degradation
exp	experimental
mod	moderately concentrated/semi-dilute
n	number average
p	polymer
R	Rouse
red	reduced
s	solvent
sp	specific
theor	theoretical
w	weight average
λ	relaxation time
0	experimental or steady state
*	critical
**	transition moderately conc–conc
+	transition dilute–moderately conc

1 Introduction

At the molecular level there are several important differences between polymeric fluids and "small-molecule" fluids. Because of these differences the flow behaviour of polymeric fluids is not similar to that of small-molecule fluids, which are satisfactorily described by Newtonian fluid dynamics. There are several salient features of macromolecular architecture that influence the flow behaviour:

- the molar masses of the constituent molecules are very high;
- the polymers are mixtures of coils with different molar masses and the rheological properties are very sensitive to the molar mass distribution;
- the polymer molecules can assume a tremendous number of configurations, even at equilibrium; then, in flow, the distribution of configurations can be greatly altered with the result that stretching and alignment of the molecules can cause the flow properties to change;
- in concentrated solutions or in melts the molecules can form a temporary entanglement network with entanglement junctions whose number can change with time in various flow situations.

These flow features are of importance in a great number of technical processes, especially for high process velocities when extremely high shear rates can be observed. For polymeric systems this can lead to a so-called non-Newtonian behaviour, i.e. the rheological material functions become dependent on the shear or elongational rate.

Some examples where the viscoelastic behaviour of polymer solutions are exploited are their use as **thickeners** in:

- the *food industry*, to improve the ability of storage, as gelling agents in fruit masses and glazes, for a better freeze-thaw stability, to adjust the mouth feeling [1–4];
- *cosmetic industry*, to improve the consistency, the homogeneity and flow properties of cosmetic products [5, 6];
- *enhanced oil recovery* as a thickener in a flooding solution [7];

as **flow improvers** in:

- *turbulent flow* (drag reduction) [8, 9];
- *water treatment and protection*, as flocculation agents [10];
- during the so-called dye-swell in the *textile industry* [11];
- *medical applications* [12], as biological fluids and in biomechanical science;

as **stabilizers** in:

- the *pharmaceutical industry*, as binding and blasting agents for pills, as solubility enhancers or as time-release preparations [4];
- the *building material industry*, as water binding agents [13], to stabilize pigments in paints, to improve the flow properties of wall-paper adhesives and coatings [14, 15];

- *industry*, as films for treating surfaces, as processing aids in mining and mineral industry, aids in textile industry, polymerization reaction aids [16];
- *agriculture*, as binders or thickeners for stick or binding fertilizers, for pesticides [17, 18], to name but a few.

In order to understand polymer solution behaviour, the samples have to be characterised with respect to their molecular configuration, their molar mass and polydispersity, the polymer concentration and the shear rate. Classical techniques of polymer characterisation (light scattering, viscometry, ultracentrifugation, etc.) yield information on the solution structure and conformation of single macromolecules, as well as on the thermodynamic interactions with the solvent. In technical concentrations the behaviour of the dissolved polymer is more complicated because additional intramolecular and intermolecular interactions between polymer segments appear.

A theoretical prediction of water-soluble polymer solutions is difficult to obtain due to their ability to build up aggregations and associations. A prediction of the viscosity yield is much easier to observe for solutions of synthetic polystyrene due to its simple solution structure. These solutions have been well characterized in other studies [19–23] concerning their chemical composition, molar mass and sample polydispersity.

In a previous paper [24] the viscoelastic properties of polyacrylamide homopolymers in solution were discussed in dependence on molar mass, concentration, solvent quality and shear rate. Considering these data a simple equation was developed for the η_0-M-c relationship.

The aim of the present paper is to report on the solution structure of polymers, to show how structure–property relationships can be derived in a simple manner, so that they can be used for technical applications. Some predictions will also be made concerning the viscous and elasticity yield as well as polymer shear stability. To demonstrate these theoretical predictions narrowly distributed polystyrene samples will mainly be used as examples.

2 Molecular Theories

The rheological behaviour of polymeric solutions is strongly influenced by the conformation of the polymer. In principle one has to deal with three different conformations, namely: (1) random coil polymers; (2) semi-flexible rod-like macromolecules; and (2) rigid rods. It is easily understood that the hydrodynamically effective volume increases in the sequence mentioned, i.e. molecules with an equal degree of polymerisation exhibit drastically larger viscosities in a rod-like conformation than as statistical coil molecules. An experimental parameter, easily determined, for the conformation of a polymer is the exponent a of the Mark-Houwink relationship [25, 26]. In the case of coiled polymers a is between 0.5 and 0.9, semi-flexible rods exhibit values between 1 and 1.3, whereas for an ideal rod the intrinsic viscosity is found to be proportional to M^2.

Different molecular theories have been established [27–32] to describe the viscoelasticity of polymeric liquids. Due to their importance, a brief survey of the different theories will be given below.

2.1 The Bead-Spring Model

First approaches at modeling the viscoelasticity of polymer solutions on the basis of a molecular theory can be traced back to Rouse [33], who derived the so-called bead-spring model for flexible coiled polymers. It is assumed that the macromolecules can be treated as threads consisting of N beads freely jointed by (N-1) springs. Furthermore, it is considered that the solution is ideally dilute, so that intermolecular interactions can be neglected.

Zimm [34] extended the bead-spring model by additionally taking hydrodynamic interactions into account. These interactions lead to changes in the medium velocity in the surroundings of each bead, by beads of the same chain. It is worth noting that neither the Rouse nor the Zimm model predicts a shear rate dependency of η. Moreover, it is assumed that the beads are jointed by an ideally Hookean spring, i.e. they obey a strictly linear force law.

2.2 The Entanglement and Reptation Concepts

2.2.1 *Linear Viscoelastic Behaviour*

In most cases polymer solutions are not ideally dilute. In fact they exhibit pronounced intermolecular interactions. First approaches dealing with this phenomenon date back to Bueche [35]. Proceeding from the fundamental work of Debye [36] he was able to show that below a critical molar mass M_w^* the zero-shear viscosity is directly proportional to M_w, whereas above this critical value η_0 is found to be proportional to ($M_w^{3.4}$) [37, 38]. This enhanced drag has been attributed to intermolecular couplings. Ferry and co-workers [39] reported that the dynamic behaviour of polymeric liquids is strongly influenced by coupling points.

For polymer melts or solutions, Graessley [40–42] has shown that for a random coil molecule with a Gaussian segment distribution and a uniform number of segments per unit volume, a shear rate dependent viscosity arises. This effect is attributed to shear-induced entanglement scission.

Introduction of the "reptation concept" by De Gennes [43] led to further essential progress. Proceeding from the notion of a reptile-like motion of the polymer chains within a tube of fixed obstacles, De Gennes [43–45], Doi [46, 47] and Edwards [48] were able to confirm Bueche's 3.4-power-law for polymer melts and concentrated polymer solution. This concept has the disadvantage that it is valid only for homogeneous solutions and no statements about flow behaviour at finite shear rates are analysed.

2.2.2
Non-Linear Elastic Behaviour

Non-linear viscoelastic flow phenomena are one of the most characteristic features of polymeric liquids. A matter of very emphasised interest is the first normal stress difference. It is a well-accepted fact that the first normal stress difference N_1 is similar to G', a measure of the amount of energy which can be stored reversibly in a viscoelastic fluid, whereas τ_{12} is considered as the portion that is dissipated as viscous flow [49–51]. For concentrated solutions Lodge's theory [52] of an elastic network also predicts normal stresses, which should be associated with the entanglement density.

These classical molecular theories may be used to illustrate good agreement with the experimental findings when describing the two extremes of concentration: ideally dilute and concentrated polymer solutions (or polymer melts). However, when they are used in the semi-dilute range, they lead to unsatisfactory results.

3
Prediction of Rheological Behaviour of Semi-Dilute Polymer Solutions at Finite Rates of Deformation

3.1
Molecular Models for Solutions

In recent years it has been shown that employing molecular models to account for the flow properties of polymer solutions represents a useful aid in solving and simplifying the practical problems that arise during the technical application of polymer solutions. Structure–property relationships were thus formulated which made it possible to describe and predict, for example, the shear viscosity as a function of polymer concentration, molar mass, and shear rate. Application of these relationships to the flow properties of polystyrene in various organic solvents [20, 22] has led to very good agreement with the mathematical predictions described by the theoretical models. However, for semi-dilute solutions it was found that in many cases no standardization of η_0 is achieved if it is plotted versus the Bueche parameter ($c \cdot M$) (see Fig. 1). Further investigations dealing with the molar mass, concentration and solvent power dependency of η_0 led to the result that the state of solution is more complex than has often been presented to date [22].

Simha [53] made the first attempts to model the transition from a dilute to a concentrated solution. He assumed that in the range from $1 \leq c \cdot [\eta] \leq 10$ the coils interpenetrate and form a homogeneous network. On the basis of the so-called scaling laws a theory has been developed which allows for the prediction of the influence of M_w, c and the solvent power on the screening length [54, 55]. This theory is founded on the presumption that above a critical concentration, c^*, the coils overlap and interpenetrate. Furthermore it is assumed that in a thermody-

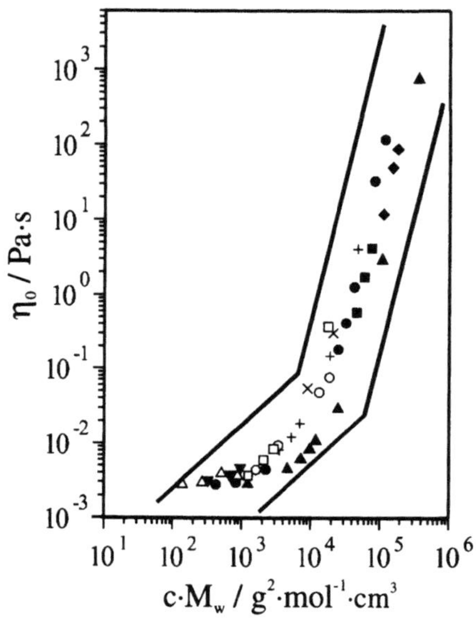

Fig. 1. Bueche plot for various narrowly distributed polystyrene samples in toluene: (Δ) M_w=48,000 g/mol; (∇) M_w=117,000 g/mol; (❑) M_w=228,000 g/mol; (◊) M_w=333,000 g/mol; (x) M_w=390,000 g/mol; (+) M_w=701,000 g/mol; (○) M_w=1,020,000 g/mol; (●) M_w=1,860,000 g/mol; (■) M_w=3,200,000 g/mol; (◆) M_w=9,150,000 g/mol; (▲) M_w=23,800,000 g/mol

namically good solvent an excluded volume is to be found. Simha [53] put forward the hypothesis that the overlap parameter ($c\cdot[\eta]$) represents a suitable means of estimating the boundaries. In the case of semi-dilute or moderately concentrated solutions, an excellent standardisation is obtained by plotting the zero-shear viscosity, expressed by the specific viscosity η_{sp}, against the overlap parameter ($c\cdot[\eta]$) (Fig. 2).

Graessley showed [40] that for a complete description of polymer fluids over the whole concentration range, i.e. from volume fraction $\phi\rightarrow 0$ to $\phi=1$, in a thermodynamically good solvent, five distinct states of solution must be taken into account. These five states of solution are:

1. ideally dilute particle solution,
2. semi-dilute particle solution,
3. semi-dilute network solution,
4. concentrated particle solution, and
5. concentrated network solution.

In a θ-solvent no semi-dilute network solution occurs, as free interpenetrability is present with overlapping. Figure 3 reproduces the individual states of solution with respect to the molar mass and the concentration [22].

The range of validity of a semi-dilute network lies between the two critical values: c* (transition between semi-dilute particle solution and semi-dilute net-

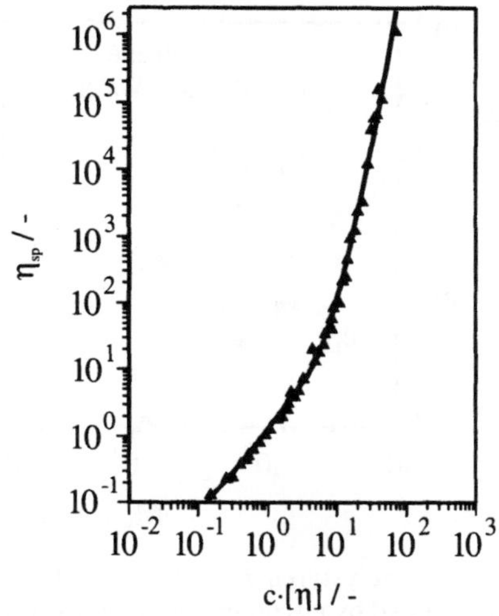

Fig. 2. Specific viscosity η_{sp} vs. the overlap parameter (c·[η]) for various narrowly distributed polystyrene samples in toluene [18]

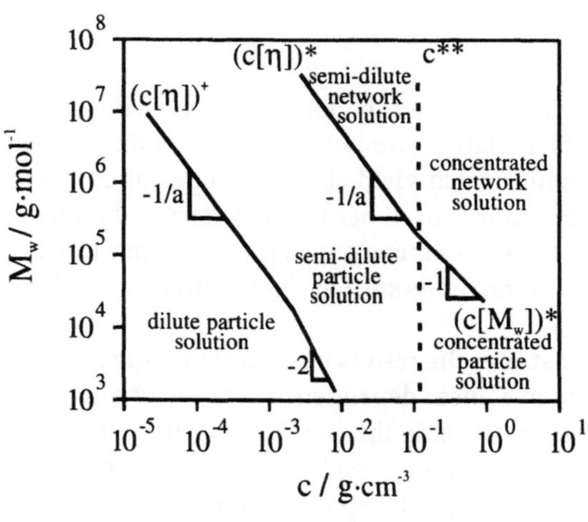

Fig. 3. Specific states of solution of narrowly distributed polystyrene in toluene as a function of the molar mass and the polymer concentration [19, 40]

work solution) and c** (transition to concentrated solution), where c* depends on the solvent and the molar mass, while c** does not.

$$c^* = \frac{K}{[\eta]} \propto M^{-a} \tag{1}$$

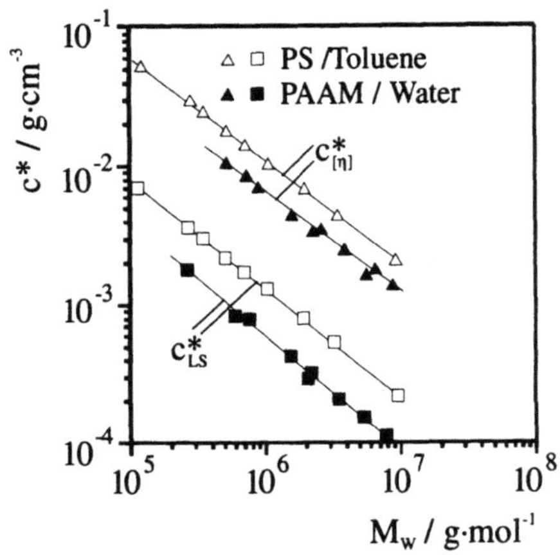

Fig. 4. Critical concentrations of polystyrene/toluene and polyacrylamide/water at 25 °C in relation to molar mass determined by viscometry and light scattering

The constant K is a function of the solvent power and increases with improving solvent quality. Furthermore c** has been shown to be a constant for a given polymer [56].

$$c^{**}=\text{const} \tag{2}$$

The parameter c* can be easily determined by using viscometric measurements, which yield a relative number, or by performing, for example, light scattering measurements, which yield absolute values of c*. Viscometric measurements are most commonly used because results are obtained quickly. However, this relative method is not equivalent to the absolute value of c^*_{LS} and only absolute determination methods such as light scattering result in correct values of c*.

Figure 4 demonstrates the results of several investigations. It can be seen that both methods lead to a linear dependence between c* and M_w but differ by a factor of ten. The reason is seen in the fact that $c^*_{[\eta]}$ depends on a model (Einstein's law), whereas c^*_{LS} gives absolute results. In both cases the geometric shape of the polymer coils are assumed to be spherical but, in accordance with the findings of Kuhn, we know that the most probable form can be best represented as a bean-like (irregularly ellipsoidal) structure.

Taking into account the relevance of the range of semi-dilute solutions (in which intermolecular interactions and entanglements are of increasing importance) for industrial applications, a more detailed picture of the interrelationships between the solution structure and the rheological properties of these solutions was needed. The nature of entanglements at concentrations above the critical value c* leads to the viscoelastic properties observable in shear flow experiments. The viscous part of the flow behaviour of a polymer in solution is usually represented by the zero-shear viscosity, η_0, which depends on the con-

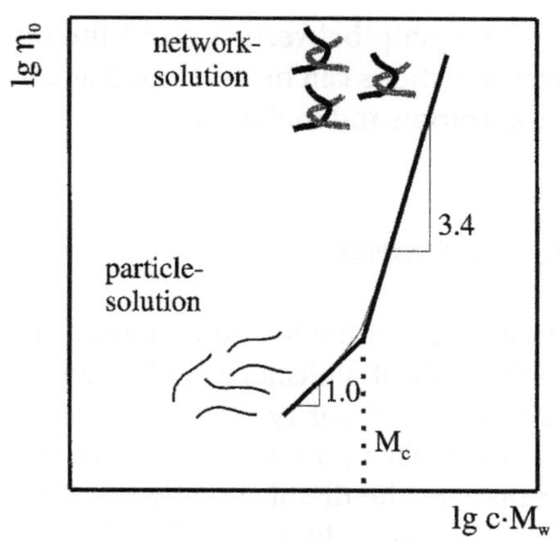

Fig. 5. Double logarithmic plot of zero-shear rate viscosity against concentration and molar mass

centration and molar mass of the dissolved polymers. It has been found that if η_0, obtainable from shear experiments at relatively low shear rates, $\dot{\gamma}$, is plotted against the molar mass in a double logarithmic plot, then the data follow two straight lines, which intersect at a critical molar mass M_c (the molar mass at which chain entanglements become important). For concentrated solutions and melts one obtains [35, 57]:

$$\eta_0 = K \cdot M^{1.0} \qquad \text{for } M < M_c \tag{3}$$

and

$$\eta_0 = K' \cdot M^{3.4} \qquad \text{for } M > M_c \tag{4}$$

as can be seen in Fig. 5. The change in M dependence occurs smoothly but over a relatively narrow range in molar mass.

In the case of melts, the linear dependence of η_0 on M_w below M_c is assumed to be due to the existence of small macromolecules that cannot build up entanglements. Once the chains are long enough to become entangled, flow becomes much more difficult because forces applied to one polymer chain are transmitted to and distributed among many other chains. The $M^{3.4}$ behaviour of melts and concentrated polymer solutions is thought to be due to the existence of entanglements between the macromolecules [57, 58]. In some cases [59–61] the slope for $M > M_c$ appears to be somewhat higher than 3.4; however, the 3.4 power is supported by data from a remarkably wide range of linear polymers [36, 57, 62].

For semi-dilute solutions, two regimes with different slopes are similarly obtained; the powers of M, however, can be lower than 1.0 and 3.4. Furthermore, the transition region from the lower to the higher slope is broadened. The critical molar mass, M_c, for polymer solutions is found to be dependent on concentration (decreasing as c increases), although in some cases the variation appears to be very small [60, 63].

On the basis of a relationship between η_{sp} and the dimensionless product $c\cdot[\eta]$, simple three-term equations can be developed to correlate the zero-shear viscosity with the concentration and molar mass.

3.1.1
Determination of η_0-M_w-c Relationships

The viscosity level in the range of the Newtonian viscosity η_0 of the flow curve can be determined on the basis of molecular models. For this, just a single point measurement in the zero-shear viscosity range is necessary, when applying the Mark–Houwink relationship. This zero-shear viscosity, η_0, depends on the concentration and molar mass of the dissolved polymer for a given solvent, pressure, temperature, molar mass distribution M_w/M_n, i.e.

$$\eta_0=f(M,c,S,p,T,M_w/M_n) \tag{5}$$

A general virial equation can be written for the specific viscosity when the zero-shear value is expressed in its specific form:

$$\eta_{sp}=B_1\cdot c\cdot[\eta]+B_2\cdot(c\cdot[\eta])^2+(c\cdot[\eta])^3+...+B_n\cdot(c\cdot[\eta])^n \tag{6}$$

where

$$\eta_{sp}=\frac{\eta_0-\eta_s}{\eta_s} \tag{7}$$

When all parameters with exponents higher than 2 are collected in one term, the following term is obtained:

$$\eta_{sp}=B_1\cdot c\cdot[\eta]+B_2\cdot(c\cdot[\eta])^2+B_n(c\cdot[\eta])^n \tag{8}$$

The Huggins equation [63] is a truncated version of Eq. (8) and is defined as follows for $B_1=1$ and $B_2=K_H$:

$$\eta_{sp}=c\cdot[\eta]+K_H(c\cdot[\eta])^2+B_n(c\cdot[\eta])^n \tag{9}$$

In an ideally dilute solution the viscosity level is governed solely by the hydrodynamically effective coil volume, i.e. the first term of Eq. (9) is the determining factor. In this concentration range intermolecular interactions are negligible. Increasing the polymer concentration, however, leads to polymer–polymer interactions, i.e. the second term in Eq. (9) is no longer negligible. From a purely viscometric point of view it seems reasonable to assume that the transition from a dilute to a moderately concentrated particle solution takes place when 1% of the viscosity level is governed by the second term of Eq. (9). The critical concentration, which is denoted as c^+, is in this case $c^+=10^{-2}/(K_H\cdot[\eta])$ and thus depends on the solvent quality and molar mass.

The fact that the second term represents to some extent intermolecular interactions readily explains the finding that in a thermodynamically poor solvent

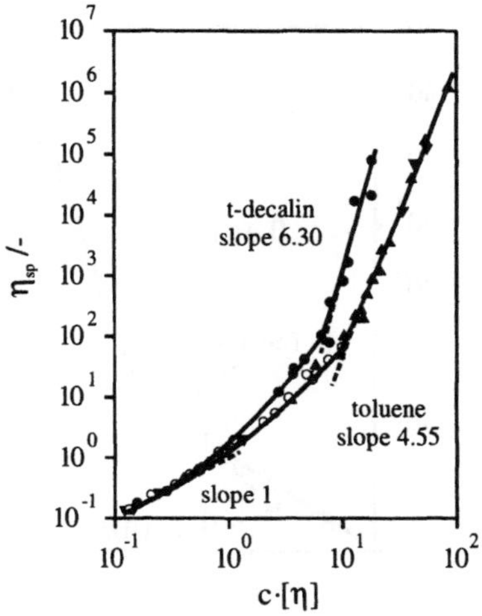

Fig. 6. Specific viscosity, η_{sp}, as a function of the product c·[η] for narrowly distributed polystyrene in toluene (good solvent) (▲) and *trans*-decalin (poor solvent) (●) at 25 °C. Experimental data for the polystyrene/toluene system at 30 °C (taken from [65]) are represented by (○).

K_H is larger and, at the same values of c·[η], the viscosity is much higher in a thermodynamically poor solvent than in a good solvent (Fig. 6).

The explanation for this is that with increasing solvent power intermolecular repulsion becomes decisive. Enhancement of the polymer concentration leads therefore to coil shrinkage [64], whereas in the limiting case of a θ-solvent no concentration-induced shrinkage is to be expected. Raising c further leads to a critical concentration, c*, at which the coils begin to overlap and interpenetrate.

A very convenient method for determining c* is provided by the η_0-M_w-c relationship. In complete analogy to Bueche, η_0 is also found to correlate in semi-dilute solutions with $M^{3.4}$. Consequently, the onset of a polymeric network is that point at which the first two terms of Eq. (9) are equal to the third term, which represents the influence of couplings on η_0.

A plot of log η_{sp} (Eq. (9)) versus log (c·[η]) results in a linear relationship. The unknown quantities B_n and n can be obtained from this linear regression (Fig. 7). A correlation of η_0 with concentration and molar mass can now be achieved using a $[\eta]=KM^a$ relationship and replacing [η] in Eq. (9) by its molar mass dependent form to give:

$$\eta_0=\eta_s(c\cdot K\cdot M^a+c^2K^2M^{2a}K_H+c^nK^nM^{na}\cdot B_n+1) \tag{10}$$

Thus, to derive a η_0-M-c relationship it is necessary to know:

1. K_H, the Huggins constant,
2. B_n and n, from a linear regression analysis at high c·[η] values of the plot log η_{sp} vs. log(c·[η]), and
3. the [η]-M relationship for the polymer-solvent system.

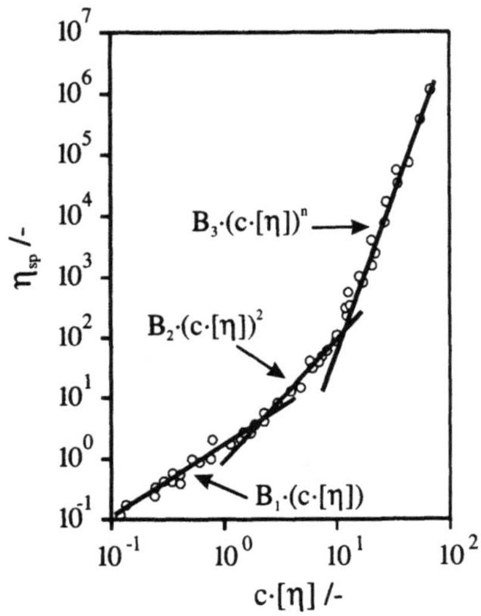

Fig. 7. Specific viscosity against c·[η] for a nonionic sample of polyacrylamide in water at 25 °C for different molar masses (M=12300–6,900,000 g/mol) and different concentrations (0.1–5 wt%)

For c·[η]>10, with the aid of the modified regression, one can obtain the value of n.

Another way to obtain the value of n is to start from Eq. (9) and consider only the part of the plot for c·[η]>10.

$$\eta_{sp}=B_n(c\cdot[\eta])^n \qquad (11)$$

Replacing the Mark–Houwink equation in Eq. (11) one obtains:

$$\log\eta_{sp}=\log(B_nK^nc^n)+a\cdot n\log M=(A)_c+a\cdot n\log M \qquad (12)$$

If $\log\eta_{sp}$ is represented against log M for different concentrations, different slopes of the representation will be obtained. If the slope is plotted against concentration, an asymptotic curve will be obtained; the intersection with the slope axis being a·n, thus the value of n can be obtained.

3.1.2
η_0-M_w-c *Relationships for Different Polymer/Solvent Systems*

The experimental zero-shear viscosities obtained for polystyrene (PS) of different molar masses (with a very narrow molar mass distribution M_w/M_n=1.06–1.30) and different concentrations in toluene and *trans*-decalin are plotted as $\log\eta_{sp}$ vs. log (c·[η]) in Fig. 6.

The data sets for each solvent are fitted by a single curve, which can be described to a good approximation by the three terms proposed in Eq. (9). Additionally, one can observe that the experimental values of PS in toluene obtained by Zakin et al. [65] at 30 °C (as plotted in Fig. 6) correspond well with the data at 25 °C. For the polystyrene solutions in toluene, the data beyond $c\cdot[\eta]=12$ were used in a linear least-square-regression analysis to estimate B_n from the intercept and n from the slope. As indicated by the correlation coefficient of 0.982, the experimental points are well described by a straight line in this range of $c\cdot[\eta]$. The $[\eta]$-M relationship for polystyrene in toluene is [19]:

$$[\eta]=8.62\cdot10^{-3}M_w^{0.736} \tag{13}$$

where the units of the intrinsic viscosity, $[\eta]$, are in $cm^3\cdot g^{-1}$. The Huggins constant K_H was estimated to be approximately 0.40. The data required for evaluating Eq. (10) are: $\eta_s=0.558$ $mPa\cdot s^{-1}$, $K=8.62\cdot10^{-3}$, $a=0.736$, $K_H=0.40$, $n=4.55$, and $B_n=2.474\cdot10^{-3}$.

The resulting η_0-M-c relationship with η_0 expressed in $mPa\cdot s^{-1}$ is then:

$$\eta_0=4.81\cdot10^{-3}\cdot c\cdot M_w^{0.736}+1.658\cdot10^{-5}\cdot c^2\cdot M_w^{1.472}+5.579\cdot10^{-13}\cdot c^{4.55}M_w^{3.35}+0.558 \tag{14}$$

The zero-shear viscosities measured in toluene solution are listed in Table 1 together with the values of η_0(theor) calculated from Eq. (14). The percentage deviation of the theoretical from the measured viscosities is given in column 8.

The values needed to calculate the η_0-M-c relationships for PS in *trans*-decalin (θ-solvent) are: $\eta_s=2.42\cdot mPa\cdot s^{-1}$, $K=6.7\cdot10^{-2}$, $a=0.52$, $K_H=0.50$, $n=6.30$, and $B_n=1.08\cdot10^{-3}$.

The resulting equation in *trans*-decalin is:

$$\eta_0=0.162\cdot c\cdot M_w^{0.52}+5.432\cdot10^{-3}\cdot c^2\cdot M_w^{1.04}+1.051\cdot10^{-10}\cdot c^{6.3}M_w^{3.28}+2.48 \tag{15}$$

The determined η_0-M-c equations [Eqs. (14) and (15)] are valid over a very wide concentration range, but they are restricted to samples having molar masses greater than approximately 20,000 g/mol. Viscosity data for lower molar masses show a more rapid increase in the η_{sp}–$(c\cdot[\eta])$ plot than the general curve, because it is assumed that the number of polymer segments is too low to form a coil.

The differences between the calculated and experimental viscosities for PS in toluene given in Table 1 are remarkably low considering the relatively large range of viscosities and values of $c\cdot[\eta]$. The viscosities of PS in *trans*-decalin calculated from Eq. (15) deviate from the experimental data more than those for the PS-toluene system (see Table 2), probably because *trans*-decalin behaves like a θ-solvent at 25 °C. The values of η_0 are presented in Table 2 and plotted in Fig. 6.

A comparison of the solution behaviour of PS in both solvents, toluene and *trans*-decalin, reveals that the limiting power of the molar mass dependence of η_0 (3.35 and 3.28, respectively) is very close to the value of 3.4 observed in highly concentrated solutions and melts. The concentration dependence of η_0, however, is clearly different in each of the solvents:

$$\eta_0\propto c^{4.55}M_w^{3.35} \text{ for PS in toluene, T=25 °C} \tag{16}$$

Table 1. Characterisation data and viscosities, η_0, of polystyrene (molar mass distribution $M_w/M_n \leq 1.3$) in toluene at 25 °C. Theoretical viscosities, η_0(theor), were calculated from Eq. (14). In the last column Δ represents the relative deviation of η_0(theor) from η_0(exp)

$M_w \cdot 10^{-5}$	[η] (cm³/g)	c (g/cm³)	c·[η]	η_{sp}	η_0(exp) (mPa·s)	η_0(theor) (mPa·s)	Δ (%)
0.48	23.9	0.0052	0.124	0.133	0.632	0.631	−0.1
		0.0100	0.233	0.266	0,706	0.705	−0.1
		0.0149	0.356	0.432	0.799	0.787	−1.5
		0.0197	0.476	0.609	0.898	0.872	−2.9
1.17	46.6	0.0049	0.228	0.242	0.693	0.696	+0.4
		0.0102	0.476	0.558	0.869	0.871	+0.2
		0.0151	0.704	0.902	1,061	1.058	−0.3
		0.0201	0.937	1.292	1.273	1.272	−0.1
2.66	83.3	0.0050	0.417	0.475	0.823	0.835	+1.5
		0.0100	0.833	1.101	1.172	1,192	+1.7
		0.0150	1.250	1.867	1.600	1.632	+21
		0.0200	1.667	2.746	2.090	2.161	+3.7
		0.1436	11.96	248.0	138.9	162.0	+16.6
3.33	99.2	0.0050	0.496	0.554	0.867	0.893	+3.0
		0.0100	0.992	1.298	1.282	1.341	+4.6
		0.0150	1.488	2.267	1.823	1.906	+4.6
		0.0200	1.984	3.411	2.461	2.600	+5.6
3.9	112.3	0.0460	5.400	30.50	17.58	11.87	−32.5
		0.1080	12.60	214.0	120.0	160.0	+33.4
7.01	172.5	0.0050	0.863	1.099	1.171	1.208	+3.2
		0.0100	1.725	2.875	2.162	2.208	+2.1
		0.0150	2.588	5.205	3.462	3.615	+4.4
		0.0200	3.450	8.592	5.352	5.556	+3.8
		0.0540	9.315	99.00	55.80	61.72	+10.6
		0.1420	24.50	3924	2190	3118	+42.4
10.19	235.5	0.0270	6.360	26.20	15.18	17.88	+17.8
		0.0360	8.480	45.50	25.95	40.33	+55.4
		0.0460	10.83	115.5	65.00	92.49	+42.3
18.6	353.3	0.0004	0.141	0.141	0.637	0.642	+0.8
		0.0008	0.283	0.262	0.704	0.734	+4.3
		0.0012	0.424	0.497	0.835	0.836	+0.1
		U016	0.565	0.723	0.931	0.947	−1.5
		0.0020	0.707	0.923	1.073	1.066	−0.6
		0.0270	9.539	113.0	63.61	67,31	+5.8
		0.0360	12.72	300.4	168.2	195.6	+21.4
		0.0465	16.43	1074	600.0	556.1	−33.6
		0.1000	35.33	43,010	24,000	16129	−39.0
		0.1434	50.66	17,0200	95,000	82,213	−25.6
32	541.0	0.0270	14.61	513.3	287.0	306.1	+6.7
		0.0360	19.48	1375	767.8	1025	+33.5
		0.0460	24.89	3780	2110	2993	+41.9
91.5	1147	0.0270	31.00	13,977	7800	8743	+12.0
		0.0360	4130	67,463	37,630	31,917	−15.2
		0.0460	5180	12,7060	70,900	96,781	+36.5

Table 1. (Continued)

$M_w \cdot 10^{-5}$	[η] (cm^3/g)	c (g/cm^3)	c·[η]	η_{sp}	η_0(exp) (mPa·s)	η_0(theor) (mPa·s)	Δ (%)
236	2245	0.0001	0.225	0.240	0.692	0.698	+0.9
		0.0004	0.898	1.185	1.219	1.262	+3.5
		0.0006	1.347	2.026	1.689	1.760	+4.4
		0.0008	1.796	3.041	2.255	2.364	+5.3
		0.0010	2.245	4.119	2.856	3.086	+8.9
		0.0020	4.490	14.73	8.780	9.317	+10.0
		0.0092	20.65	2741	1530	1630	+4.5
		0.0098	22.00	2866	1600	2149	+30.6
		0.0360	80.82	12,54000	69,9732	754,913	+7.9

Table 2. Characterisation data and viscosities, η_0, of polystyrene (molar mass distribution $M_w/M_n \leq 1.3$) in *trans*-decalin at 25 °C. Theoretical viscosities, η_0(theor) were calculated from Eq. (15). In the last column Δ represents the relative deviation of η_0(theor) from η_0(exp)

$M_w \cdot 10^{-5}$	[η] (cm^3/g)	c (g/cm^3)	c·[η]	η_{sp}	η_0(exp) (mPa·s)	η_0(theor) (mPa·s)	Δ (%)
2.66	44.4	0.0465	2.06	5.8	16.5	12.8	−22.3
		0.0978	4.34	41.6	103.0	64.2	−37.7
		0.1431	6.35	123.0	300.0	379.7	+26.6
		0.2172	9.64	949.4	2300.0	4477.5	+94.7
3.9	54.1	0.0500	2.71	10.2	27.0	19.3	−28.6
		0.1090	5.90	85.8	210.0	256.6	+22.2
7.01	73.4	0.0465	3.43	25.4	64.0	31.1	−51.4
		0.0974	7.18	396.0	960.0	747.6	−22.1
		0.1418	10.40	1735.4	4200.0	7256.5	+72.8
		0.2172	16.02	21900.0	53000.0	104526.4	+97.2
7.42	75.6	0.0464	3.50	21.7	55.0	33.3	−39.4
		0.2172	16.40	82643.6	200000.0	125898.0	−37.1
18.6	122.0	0.0464	5.70	346.0	M.O	207.8	−75.3
		0.1000	12.20	16528.0	40000.0	19514.0	−51.2
		0.1432	17.50	85536.2	207000.0	180495.0	−12.8

$$\eta_0 \propto c^{6.30} M_w^{3.28} \text{ for PS in } trans\text{-decalin, T=25 °C} \quad (17)$$

The zero-shear viscosities of PS in the poorer solvent *trans*-decalin depend on a higher exponent of c than in the better solvent toluene. Although it has been assumed in the literature that $\eta_0 \propto c^4 \cdot M^{3.5}$ [66], experimental investigations show higher exponents for c [67]. In the high molar mass region and at high concentration, the $\eta_0 \propto M^{3.4}$ relationship was established [62].

The [η]-M relationships of polystyrene ($2000<M_w<24 \cdot 10^6$) in different solvents at 25 °C and under θ-conditions (cyclohexane at 34.5 °C) are summarized in Table 3.

Table 3. Mark–Houwink relationships for polystyrene in different solvents at 25 °C as compared with θ-conditions

Parameters	Toluene	*cis*-Decalin	*trans*-Decalin	Cyclohexane
a	0.736	0.574	0.520	0.500
$K_{[\eta]}$ (cm^3/g)	0.00862	0.0400	0.0670	0.0828

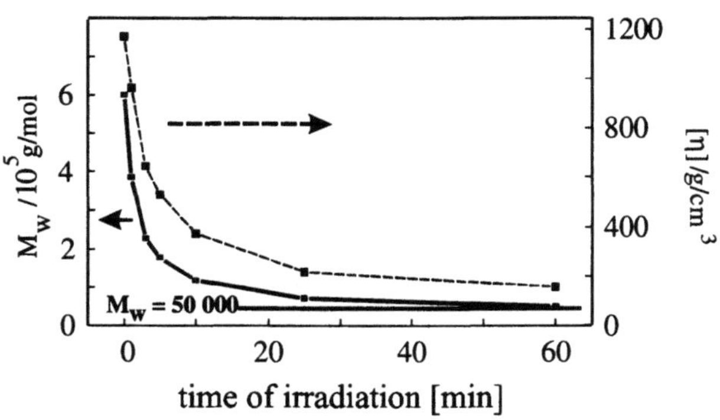

Fig. 8. Behaviour of molar mass vs. time of irradiation for methylhydroxyethylcellulose exposed to ultrasound degradation

For a correct determination of the Mark–Houwink relationship for a homologous series of polymers, it is very useful to use ultrasonic irradiation. Due to its good reproducibility, it is the best means of breaking down the molar mass and it has the advantage that it results in a definite limiting molar mass (Fig. 8). Ultrasonic degradation has the advantage that chain scission occurs only in the middle of the macromolecule, so that no low molar mass degradation products or even monomers are generated. The limiting molar mass ensures that ultrasonic degradation produces representative fragments of the original polymer. When carrying out ultrasonic degradation, it is essential to avoid any radical depolymerisation or thermal decomposition.

In Table 4 some η_0-M-c relationships for other polymer/solvent systems [68–70] are presented. For polyacrylamide (nonionic sample) in water two relationships were obtained. The values are slightly different, which was caused by one apparatus being more complex than the other. A relationship for a cationic polymer [poly(acrylamide-co-trimethylammonium methacrylate chloride)] is also shown in Table 4. It is important to notice that in dimethyl sulfoxide, schizophyllan (nonionic polymer) has a coil structure. A η_0-M-c relationship for a set of carboxymethylcelluloses (CMC) with the degree of substitution varying between 0.71 to 2.95 and the partial degree of substitution raised uniformly at positions C2, C3 and C6 is also presented. Surprisingly, the viscosity for all the CMC samples in 0.01 M NaCl at 25 °C can be predicted by a single η_0-M-c equation, independent of the degree of substitution. The accuracy is limited for these

Table 4. η_0-M-c relationships for some polymer/solvent systems

Polymer system	η_0-M-c relationship (η_0/mPa·s)
Polystyrene/toluene	$\eta_0 = 0.558 + 4.81\cdot10^{-3}\cdot c\cdot M_w^{0.74} + 1.658\cdot10^{-5}\cdot c^2\cdot M_w^{1.47} + 5.579\cdot10^{-13}\cdot c^{4.55}\cdot M_w^{3.35}$
Polystyrene/t-decaline	$\eta_0 = 2.48 + 0.162\cdot c\cdot M_w^{0.52} + 5.432\cdot10^{-3}\cdot c^2\cdot M_w^{1.04} + 1.051\cdot10^{-10}\cdot c^{6.3}\cdot M_w^{3.28}$
Polystyrene/ n-butylbenzene	$\eta_0 = 1.30 + 2.13\cdot10^{-2}\cdot c\cdot M_w^{0.68} + 1.61\cdot10^{-4}\cdot c^2\cdot M_w^{1.37} + 1.73\cdot10^{-14}\cdot c^{5.44}\cdot M_w^{3.28}$
Polystyrene/cis-decaline	$\eta_0 = 2.93 + 0.117\cdot c\cdot M_w^{0.574} + 2.34\cdot10^{-3}\cdot c^2\cdot M_w^{1.148} + 1.85\cdot10^{-11}\cdot c^{5.92}\cdot M_w^{3.28}$
Poly(methyl methacrylate)/toluene	$\eta_0 = 0.558 + 3.3\cdot10^{-3}\cdot c\cdot M_w^{0.74} + 6.2\cdot10^{-6}\cdot c^2\cdot M_w^{1.48} + 1.1\cdot10^{-10}\cdot c^{4.06}\cdot M_w^{3.0}$
Polyacrylamide/water	$\eta_0 = 0.891 + 8.01\cdot10^{-3}\cdot c\cdot M_w^{0.76} + 3.56\cdot10^{-5}\cdot c^2\cdot M_w^{1.52} + 1.77\cdot10^{-14}\cdot c^{5.23}\cdot M_w^{3.95}$
Poly(acrylamide-co-trimethylammonium-methacrylate chloride)/ 1M NaCl	$\eta_0 = 0.985 + 2.19\cdot10^{-2}\cdot c\cdot M_w^{0.65} + 1.79\cdot10^{-4}\cdot c^2\cdot M_w^{1.3} + 1.18\cdot10^{-11}\cdot c^{4.68}\cdot M_w^{3.04}$
Schizophyllan/ dimethylsulfoxide	$\eta_0 = 1.81 + 4.20\cdot10^{-2}\cdot c\cdot M_w^{0.69} + 5.38\cdot10^{-4}\cdot c^2\cdot M_w^{1.38} + 1.16\cdot10^{-41}\cdot c^{17.9}\cdot M_w^{12.3}$
Carboxymethylcellulose/ 0.01M NaCl	$\eta_0 = 0.891 + 7.82\cdot10^{-3}\cdot c\cdot M_w^{0.93} + 1.77\cdot10^{-5}\cdot c^2\cdot M_w^{1.86} + 4.22\cdot10^{-12}\cdot c^{4.09}\cdot M_w^{3.80}$
(1,3)(1,4)-β-glucane/water	$\eta_0 = 0.891 + 5.38\cdot10^{-2}\cdot c\cdot M_w^{0.71} + 2.28\cdot10^{-3}\cdot c^2\cdot M_w^{1.42} + 2.41\cdot10^{-12}\cdot c^{5.18}\cdot M_w^{3.68}$
	η_0-[η]-c relationship
Polyurethane/N,N-Dimethylacetamide	$\eta_0 = 0.942 + 0.942\cdot c\cdot[\eta] + 5.95\cdot10^{-1}\cdot(c\cdot[\eta])^2 + 1.11\cdot10^{-2}\cdot(c\cdot[\eta])^{5.17}$
Xanthane/0.1M NaCl	$\eta_0 = 0.9 + 0.9\cdot c\cdot[\eta] + 5.07\cdot10^{-1}\cdot(c\cdot[\eta])^2 + 1.34\cdot10^{-1}\cdot(c\cdot[\eta])^{3.50}$
Xanthane/water	$\eta_0 = 0.891 + 0.891\cdot c\cdot[\eta] + 5.52\cdot10^{-1}\cdot(c\cdot[\eta])^2 + 4.66\cdot10^{-2}\cdot(c\cdot[\eta])^{3.75}$

Table 4. (Continued)

Polymer system	η_0-M-c relationship (η_0/mPa·s)
Schizophyllan/water	$\eta_0 = 0.891 + 0.891 \cdot c \cdot [\eta] + 3.74 \cdot 10^{-1} \cdot (c \cdot [\eta])^2 + 3.65 \cdot 10^{-4} \cdot (c \cdot [\eta])^{5.03}$
Hydroxypropylcellulose/ water	$\eta_0 = 0.891 + 0.891 \cdot c \cdot [\eta] + 6.11 \cdot 10^{-1} \cdot (c \cdot [\eta])^2 + 1.28 \cdot 10^{-1} \cdot (c \cdot [\eta])^{4.24}$

polymers by the supposition that aggregates or associations can appear. A η_0-M-c relationship for a (1,3)(1,4)-β-glucan/water system was obtained. Almost the same dependency was obtained by Robinson [71] for an oat β-glucan/water system; the plot of η_{sp}=f(c·[η]) having slopes of 1.08, 1.8 and 3.90, respectively. These relationships can be a good help for technical applications.

In order to evaluate the viscosity of a polymeric liquid at finite rates of deformation, two parameters must be determined, i.e. (i) the critical shear rate $\dot{\gamma}$ ($\dot{\gamma}$=1/λ) at which η becomes a function of the of deformation, and (ii) the slope in the linear range of the flow curve.

3.2 η–M_w–c–$\dot{\gamma}$ Relationship

The influence of the molar mass and concentration above the zero-shear viscosity has been described. In the following sections the influence of these parameters in the region of Newtonian and non-Newtonian regimes will be discussed.

3.2.1 *Concentration Dependence*

Polymers in solution or as melts exhibit a shear rate dependent viscosity above a critical shear rate, $\dot{\gamma}_{crit}$. The region in which the viscosity is a decreasing function of shear rate is called the non-Newtonian or power-law region. As the concentration increases, for constant molar mass, the value of $\dot{\gamma}_{crit}$ is shifted to lower shear rates. Below $\dot{\gamma}_{crit}$ the solution viscosity is independent of shear rate and is called the zero-shear viscosity, η_0. Flow curves (plots of log η vs. log $\dot{\gamma}$) for a very high molar mass polystyrene in toluene at various concentrations are presented in Fig. 9. The transition from the shear-rate independent to the shear-rate dependent viscosity occurs over a relatively small region due to the narrow molar mass distribution of the PS sample.

3.2.2 *Molar Mass Dependence*

Qualitatively, the same behaviour is observed for the flow curves at a fixed polymer concentration with various molar masses (Fig. 10). The shear rate depend-

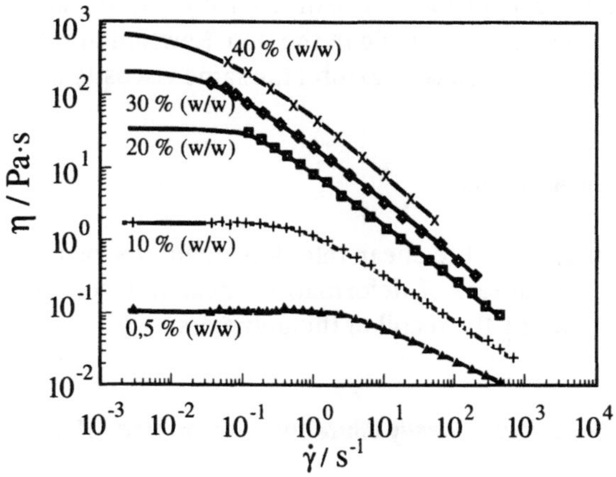

Fig. 9. Viscosity h vs. shear rate $\dot{\gamma}$ for polystyrene (MW=23.6◊106) in toluene for various concentrations at 25 °C

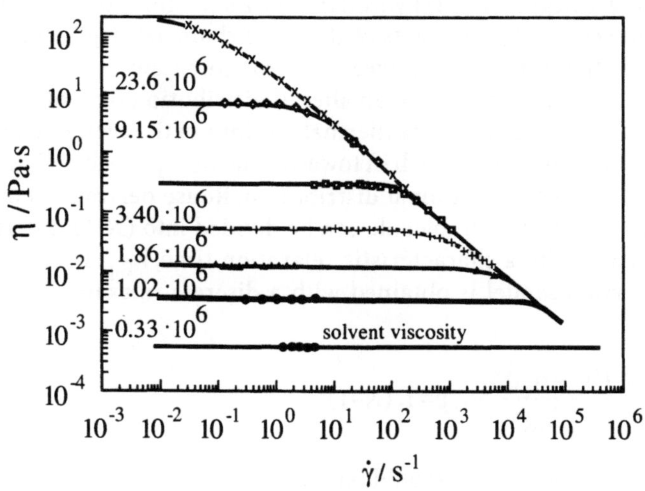

Fig. 10. Viscosity vs. shear rate for polystyrene samples of various molar masses in toluene at 25 °C

ence of the viscosity begins at a lower value of $\dot{\gamma}$ than the molar mass increases. However, for a very high polystyrene molar mass and a solution concentration of 3 wt%, the viscosities in the non-Newtonian region become equal for all the samples with various molar masses and follow a straight line with a slope of –0.83. This slope is in accordance with the theory of Graessley (–0.82) [57] and other experimental data [72, 73].

The fact that viscosity is independent of molar mass at high shear rates is of fundamental importance, since it follows that it is impossible to distinguish between different samples if the viscosity is measured in the power-law region.

Moreover, a further conclusion can be drawn from Fig. 10, namely, that the evaluation of molar mass is impossible in the non-Newtonian region because samples with different molar mass all exhibit the same viscosity at a given shear rate.

3.3
Characteristic Relaxation Time

For a precise analysis of the shear rate dependent viscosity it is necessary to know at which critical rate of deformation shear-induced disturbance can no longer be leveled out by the recoil of the polymers.

3.3.1
Relaxation Time Behaviour in Ideally Dilute and Concentrated Solutions

First approaches to approximating the relaxation time on the basis of molecular parameters can be traced back to Rouse [33]. The model is based on a number of boundary assumptions: (1) the solution is ideally dilute, i.e. intermolecular interactions are negligible; (2) hydrodynamic interactions due to disturbance of the medium velocity by segments of the same chain are negligible; and (3) the connector tension F(r) obeys an ideal Hookean force law.

The velocity gradient leads to an altered distribution of configuration. This distortion is in opposition to the thermal motions of the segments, which cause the configuration of the coil to drift towards the most probable distribution, i.e. the equilibrium's configurational distribution. Rouse derivations confirm that the motions of the macromolecule can be divided into (N-1) different modes, each associated with a characteristic relaxation time, $\lambda_{R,p}$. In this case, a generalised Maxwell model is obtained with a discrete relaxation time distribution.

$$\lambda_{R,p} = \frac{6}{\pi^2} \frac{(\eta_0 - \eta_s) M_w}{p^2 c R T}, \; p=1, (N-1) \tag{18}$$

The longest mode (p=1) should be identical to the motion of the chain. The fundamental correctness of the model for dilute solutions has been shown by Ferry [74]. Ferry and co-workers [39, 75] have shown that, in concentrated solutions, the formation of a polymeric network leads to a shift of the characteristic relaxation time λ_0 ($\lambda_0 = 1/\dot{\gamma}_{crit}$; i.e. the critical shear rate where η becomes a function of $\dot{\gamma}$). It has been proposed that this time constant is related to the motion of the polymeric chain between two coupling points.

For concentrated polystyrene solutions (c>c**) in *n*-butylbenzene, Graessley and co-workers [36] observed that the shift factor h^{conc} depends on the number of entanglements per macromolecule E.

$$\lambda_0^{conc} = h^{conc} \cdot \lambda_R \tag{19}$$

$$h^{conc} = f(E) = K_1 \cdot (1 + K_2 \cdot E)^{-1} \tag{20}$$

For highly concentrated solutions, Eq. (20) can be simplified, due to the fact that in such a case the plateau modulus G_p' is M_w independent. Under these conditions it is clear that E is solely a function of M_w.

$$E = \frac{M_w}{M_e} - 1 = \text{const.} \cdot G_p' \cdot M_w - 1 = \text{const.} \cdot M_w - 1 \tag{21}$$

By substitution of E [Eq. (21)] in Eq. (20), on the assumption that $(K_2 \cdot \text{const.}' \cdot M_w) >> (1 - K_2)$, one obtains:

$$h^{conc} \cong \text{const.} \cdot M_w^{-1} \tag{22}$$

Substitution of $(\eta_0 - \eta_s)$ in Eq. (18) by Bueche's 3.4-power-law [46, 75] $(\eta_0 \propto M_w^{3.4})$ leads to the following dependency:

$$\lambda_0^{conc} \propto \lambda_R \cdot M_w^{-1} \propto M_w^{3.4} \tag{23}$$

However, these results are only valid for freely interpenetrating coils, a molar mass independent screening length ξ and a uniform segment density.

3.3.2
Relaxation Time Behaviour of Moderately Concentrated Polymeric Solutions

In semi-dilute solutions, the Rouse theory fails to predict the relaxation time behaviour of the polymeric fluids. This fact is shown in Fig. 11 where the reduced viscosity is plotted against the product ($\dot{\gamma} \cdot \lambda_R$). For correctly calculated values of λ_0 a satisfactory standardisation should be obtained independently of the molar mass and concentration of the sample.

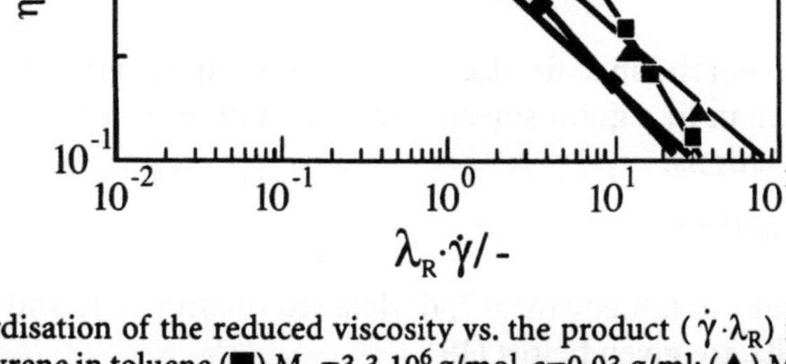

Fig. 11. Standardisation of the reduced viscosity vs. the product ($\dot{\gamma} \cdot \lambda_R$) for narrowly distributed polystyrene in toluene (■) M_w=3.3·10^6 g/mol, c=0.03 g/ml; (▲) M_w=3.3·10^6 g/mol, c=0.04 g/ml; (◆) M_w=9.2·10^6 g/mol, c=0.03 g/ml; (▼) M_w=23·10^6 g/mol, c=0.05 g/ml

The range of semi-dilute network solutions is characterised by: (1) polymer–polymer interactions which lead to a coil shrinkage; (2) each blob acts as individual unit with both hydrodynamic and excluded volume effects; and (3) for blobs in the same chain all interactions are screened out (the word blob denotes the portion of chain between two entanglements points). In this concentration range the flow characteristics and therefore also the relaxation time behaviour are not solely governed by the molar mass of the sample and its concentration, but also by the thermodynamic quality of the solvent. This leads to a shift factor, h^{mod}, that is a function of the molar mass, concentration and solvent power.

$$h^{mod} = \frac{\text{const.} \cdot [\eta]}{K_\eta \cdot M_w \cdot c^{1/(a-1)}} \tag{24}$$

Substitution of the term (η_0–η_s) by the η_0-M_w-c relationship [Eq. (10)] leads to a general form of the λ_0-M_w-c relationship:

$$\begin{aligned} \lambda_0^{mod} &= \text{const.} \cdot \frac{[\eta](\eta_0 - \eta_s)}{R \cdot T \cdot c^{1/a}} = \\ &= \text{const.} \left[\sum_{i=2}^{3} B_i (c \cdot [\eta])^i + B_n (c[\eta])^{n+1} \right] c^{-[1+1/a]} \end{aligned} \tag{25}$$

It is worth noting that for the limiting case of a=1, the λ_0-M_w-c relationship exhibits the same proportionalities as obtained for the Rouse equation [Eq. (18)]. On the basis of Eq. (25) it is possible to obtain a standardisation of the relaxation time if the product $\lambda_0 \cdot c^{1+1/a}$ is plotted against the overlap parameter, $c \cdot [\eta]$, as shown in Fig. 12.

For high values of $c \cdot [\eta]$ the final term in Eq. (25) becomes the determining factor.

$$\lambda_0^{mod} \cdot c^{1+1/a} \propto (c \cdot [\eta])^{n+1} \tag{26}$$

It has been proved [20] that the exponent n of the η_0-M_w-c relationship is a function of the solvent quality:

$$n = 3.4/a \tag{27}$$

i.e. in a poor solvent the specific viscosity increases more rapidly with increasing values of $c \cdot [\eta]$ than in a good solvent. Substitution of n in Eq. (27) leads to the following correlations:

$$\lambda_0^{mod} \propto c^{2.4/a} [\eta]^{3.4/(a+1)} \tag{28}$$

For polystyrene in toluene (a=0.736) detailed theoretically and experimentally derived results are given below [19].

$$\lambda_0^{mod}{}_{,theor} \propto c^{3.26} [\eta]^{5.62} \tag{29}$$

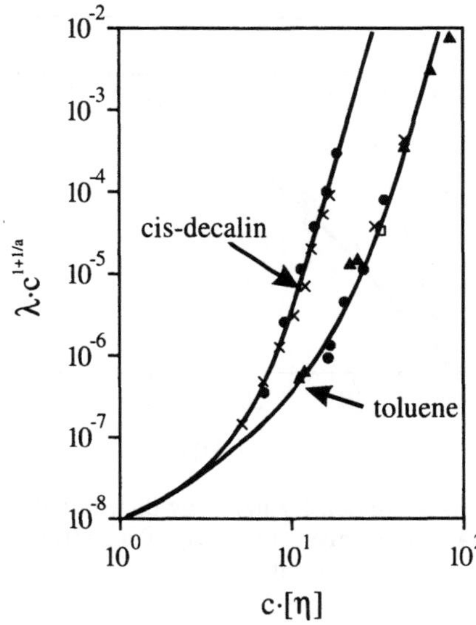

Fig. 12. Relaxation time standardisation for narrowly distributed polystyrene in toluene and *cis*-decalin. (x) M_w=1.86·10^6 g/mol; (●) M_w=3.20·10^6 g/mol; (❑) M_w=9.15·10^6 g/mol; (▲) M_w=23.66·10^6 g/mol

$$\lambda_0{}^{mod}{}_{,exp} \propto c^{3.02}[\eta]^{5.37} \tag{30}$$

The total λ_0-M_w-c relationship for polystyrene at 25 °C in toluene is:

$$\lambda_0 = 2.2 \cdot 10^{-10} \cdot c^{-0.36} \cdot [\eta]^2 + 8.8 \cdot 10^{-11} \cdot c^{0.64} \cdot [\eta]^3 + 5.4 \cdot 10^{-13} \cdot c^{3.02} \cdot [\eta]^{5.37} \tag{31}$$

Using the statements formulated above it is now possible to determine the relaxation time for every state of solution.

3.4 Slope of the Flow Curve in the Power-Law Region

The flow characteristics at finite shear of narrowly distributed melts have been investigated very thoroughly by Stratton [76]. He reported that in the power-law region, a slope n of –0.82 is obtained. For concentrated solutions Graessley [41] had theoretically derived an approximate identical value of –0.818. However, these approaches are only valid for homogeneous solutions with a uniform segment density. For non-homogeneous solutions, a pronounced concentration dependency of the slope in the linear region of the flow curve has been observed experimentally. In this case a limiting slope of –0.83 was measured for high values of c·[η] [22]. This is, within the margins of error, commensurate with both Stratton's and Graessley's data. Yet in less concentrated solutions [22] a pronounced dependency between the slope n (dlog η/dlog $\dot{\gamma}$) and the overlap pa-

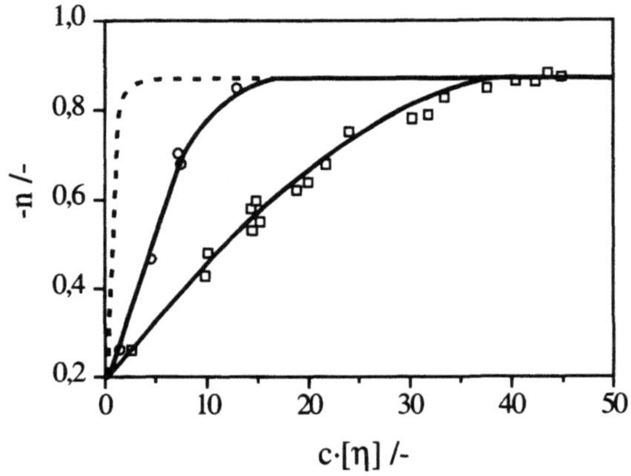

Fig. 13. Slope of the power-law region for narrowly distributed polystyrene in (○) *trans*-decalin (thermodynamically poor solvent) and (□) toluene (thermodynamically good solvent). (–) theoretical course for a θ-solvent

rameter was observed. As shown in Fig. 13, n is not solely a function of c·[η] but depends also on the solvent quality.

Analysis of the experimental data leads to the assumption that n is a measure of the homogeneity of a solution, i.e. in a θ-solvent the slope should be equal to –0.83 for $c\cdot[\eta]\geq 0.7$, due to the fact that there is unhindered interpenetration and therefore a homogeneous segment density. However, it is worth mentioning that it is impossible to verify this fact by simple shear flow experiments. As Wissbrunn [77] showed, thermodynamic interactions can be seriously influenced by a velocity field. Increasing the velocity can lead to a flocculation of the polymer which is coupled with a phase transition at, or in the region of, the θ-point. In a thermodynamically good solvent the limiting slope of –0.83 should be attained for $c>c^{**}$ (e.g. for polystyrene the transition to a concentrated, homogeneous solution occurs at $c^{**}=0.118$ g/ml) or for large values of c·[η]. The interdependence between n and c·[η] can be expressed using a simple exponential approach.

$$\log(n+0.83)=K_1-K_2\cdot c\cdot[\eta] \tag{32}$$

The constants K_1 and K_2 must be evaluated experimentally as shown in Fig. 14.

However, the fact that the concentration has a more pronounced influence on the slope in the power-law region of the flow curve than M_w is often overlooked. Variation of the concentration (M_w=const.) therefore leads to a drastic change in the slope, whereas a variation of the molar mass (c=const.) gives rise to a far less substantial change in n (Fig. 15). This fact is easily understood from Eq. (32). Log n increases with increasing molar mass, giving the slope a ($\log n \propto M^a$; a<1; c=const.), whereas log n is directly proportional to c ($\log n \propto c$; M_w=const.).

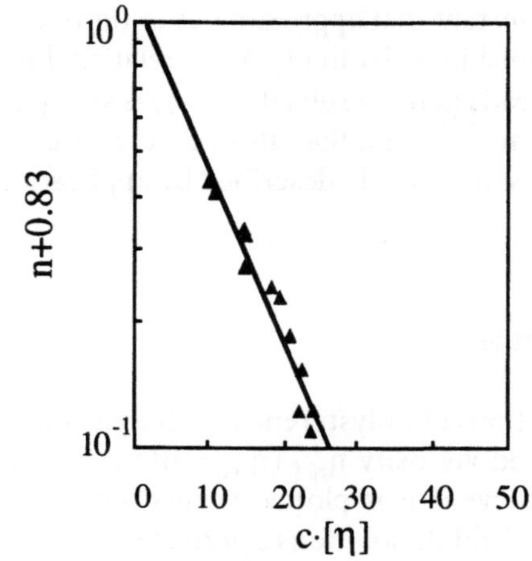

Fig. 14. Slope in the power-law region of the flow curve as a function of the overlap parameter for narrowly distributed polystyrene in toluene

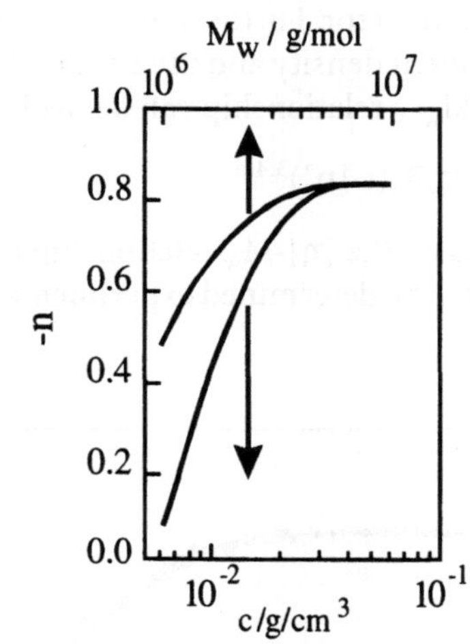

Fig. 15. Influence of molar mass and concentration on the slope of the flow curve for narrowly distributed polystyrene in toluene (n=f(c) for $M_w=1\cdot10^7$ g/mol; n=f(M_w) for c= 0.06 g/ml)

It is possible to approximate the shear rate dependent viscosity at any rate of deformation ($\dot{\gamma}\geq\dot{\gamma}_{crit}$), to such an extent that degradation may be neglected.

$$\frac{(\eta(\dot{\gamma})-\eta_s)}{(\eta_0-\eta_s)}=\text{const.}\cdot(\dot{\gamma}/\dot{\gamma}_{cr})^n=\text{const.}\cdot(\dot{\gamma}\cdot\lambda_0)^n \tag{33}$$

$$\eta(\dot{\gamma})=\eta_s+K\cdot(\dot{\gamma}\cdot\lambda_0)^n\cdot(\eta_0-\eta_s) \tag{34}$$

In Eq. (34) K is a constant that approximates to unity. Substitution of $(\eta_0-\eta_s)$ by the η_0-M_w-c relationship, λ_0 by the λ_0-M_w-c relationship and the exponent n, as mentioned above, leads to the results that $\eta(\dot{\gamma})$ is solely a function of the overlap parameter, the term $c^{1+1/a}$ and the rate of deformation, i.e. the shear rate dependence of η can be satisfactorily described by application of a three-parameter approach.

3.5 Flow Curve Standardisation

For concentrated solutions of polystyrene in *n*-butylbenzene, Graessley [40] has shown that the reduced viscosity η_{red} ($\eta_{red}=(\eta(\dot{\gamma})-\eta_s)/(\eta_0-\eta_s)$) can be represented on a master curve if it is plotted versus the reduced shear rate β ($\beta=\dot{\gamma}/\dot{\gamma}_{crit}=\dot{\gamma}\cdot\lambda_0$). For semi-dilute solutions a perfect master curve is obtained if β is plotted versus a slope corrected for reduced viscosity, η_{corr}, as shown in Fig. 16.

In cases of relatively high values of $c\cdot[\eta]$, ($c\cdot[\eta]\geq 15$ in poor solvents and $c\cdot[\eta]\geq 25$ in good solvents), the influence of the homogeneity of the slope in the linear region of the flow curve is negligible and a slope of −0.83 can be assumed without risking too great an error. Furthermore, the viscosity in this region is governed by the entanglement density and the excluded volume. In this case, the first two terms of the η_0-M_w-c relationship can be neglected.

$$(\eta 0-\eta_s)\cong\eta_s B_n\cdot(c\cdot[\eta]^n=\eta_s B_n(c\cdot[\eta])^{3.4/a} \quad (35)$$

In a great number of cases the $[\eta]$-M_w relationship can be taken from handbooks, so that only B_n has to be determined experimentally. This can be achieved

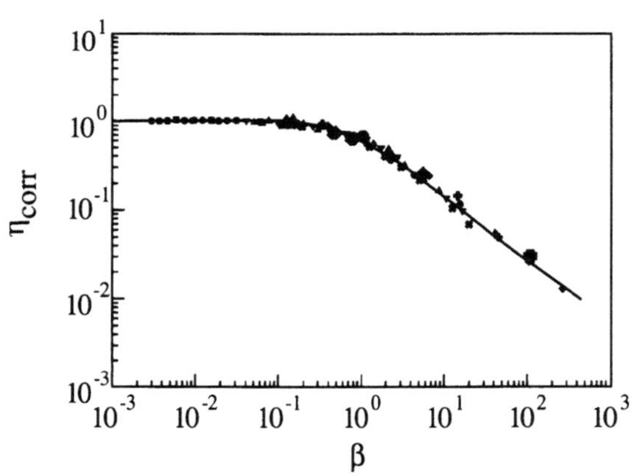

Fig. 16. Flow curve standardisation for moderately concentrated solutions of polystyrene in various solvents: (●) $M_w=3.2\cdot10^6$ g/mol, c=0.100 g/ml, *n*-butylbenzene at 50 °C; (+) $M_w=23\cdot10^6$ g/mol, c=0.010 g/ml, toluene at 25 °C; (x) $M_w=3.2\cdot10^6$ g/mol, c=0.080 g/ml, *n*-butylbenzene at 25 °C; (◆) $M_w=9.2\cdot10^6$ g/mol, c=0.025 g/ml, toluene at 25 °C; (■) $M_w=6.3\cdot10^6$ g/mol, c=0.080 g/ml, toluene at 25 °C; (▲) $M_w=3.2\cdot10^6$ g/mol, c=0.080 g/ml, toluene at 25 °C; (▼) $M_w=3.2\cdot10^6$ g/mol, c=0.040 g/ml, toluene at 25 °C

by a simple viscosity measurement. Using this approach the λ_0-M_w-c relationship [78] can also be greatly simplified by substitution of $(\eta_0-\eta_s)$ in Eq. (36):

$$\lambda_0 = \text{const.} \frac{[\eta](\eta_0 - \eta_s)}{RTc^{1/a}} \cong K_\lambda \cdot \frac{c^{2.4/a} \cdot [\eta]^{\frac{3.4}{a}+1}}{T} \tag{36}$$

As in Eq. (35), the constant K_λ in Eq. (36) can be evaluated by determination of $\dot{\gamma}_{crit}$ in one example. For the shear rate dependent viscosity, it then follows that $(\eta(\dot{\gamma}) >> \eta_s)$:

$$\begin{aligned} \eta(\dot{\gamma}) &\cong \left[\frac{K_\lambda \cdot c^{2.4/a} \cdot [\eta]^{\frac{3.4}{a}+1} \cdot \dot{\gamma}}{T} \right]^{-0.83} \cdot \eta_s \cdot B_n \cdot (c \cdot [\eta]^{\frac{3.4}{a}}) \cong \\ &\cong B_n \cdot c^{1.4/a} \cdot [\eta]^{\frac{0.58}{a}-0.83} \cdot \left[\frac{K_\lambda \cdot \dot{\gamma}}{T} \right]^{-0.83} \cdot \eta_s \end{aligned} \tag{37}$$

3.6
The Second Newtonian Region

For high rates of deformation, a second shear rate independent region of the viscosity η_∞ (the second Newtonian region) has been predicted in a great number of cases [79, 80]. Following the entanglement concept the transition can be understood on a molecular level by a decreasing concentration of entanglement points with increasing rate of shear. Thus, at very high shear rates the molecules are finally extended to a maximum and therefore an upper limiting viscosity value should be reached. Phenomenological theories [81] as well as simple empirical equations (extended power-law type) [82] make use of a definite value η_∞ in modelling the complete flow curve.

In a set of viscometric measurements the behaviour of different concentrations of polystyrene in decalin solutions up to extreme high shear rates has been investigated [83] (Fig. 17), in an attempt to evaluate the upper limiting viscosity η_∞. As can clearly be seen, the solution viscosity at the highest shear rates is only two-thirds of a decade above the solvent viscosity. Extrapolation of the power-law slope to higher shear rates in each case intersects with the solvent viscosity at $5 \cdot 10^6\ s^{-1}$. η_∞ is not reached and therefore is not a deviation from a straight slope, i.e. a power-law region. Flow behaviour similar to that of PS solutions was found in other studies and η_∞ was not detectable [72]. The main experimental problems which arise at very high shear rates are chain scission and shear heating. One can conclude that η_∞ is experimentally not readily accessible. It may be even be impossible to maintain the laminar flow conditions of viscoelastic fluids of coiled polymers at high shear rates and thus to determine η_∞.

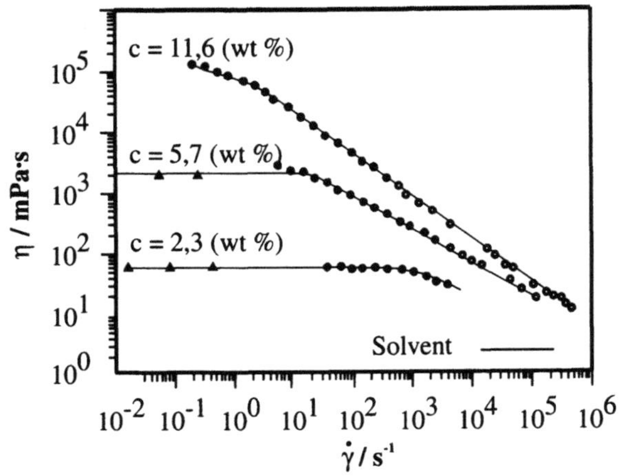

Fig. 17. Flow curve for various concentrations of polystyrene (M_w=2·10^6 g/mol) in decalin solutions at 25 °C obtained with a low-shear viscometer (Δ), a mechanical spectrometer (●) and a high-shear concentric-cylinder viscometer (O)

3.7 Criteria for Shear Stability

In order to obtain solutions with the desired flow properties, shear-induced degradation should be avoided. From mechanical degradation experiments it has been shown that chain scission occurs when all coupling points are loose and the discrete chains are subjected to the velocity field. Simple considerations lead to the assumption that this is obtained when $\eta_{sp}(\dot{\gamma})$ is equal to $\eta_{sp}(c\cdot[\eta])$ (Fig. 18). The critical shear rate can then easily be evaluated [22].

$$\frac{\eta_{sp,deg}}{\eta_{sp,0}} = (\dot{\gamma}_{deg}\cdot\lambda)^n \tag{38}$$

Substitution of $\eta_{sp,deg}$ by $c\cdot[\eta]$ and $\eta_{sp,0}$ by the η_0-M_w-c relationship (Eq. (10)) leads to the following equation:

$$\dot{\gamma}_{deg} = \left[1 + K_H\cdot c\cdot[\eta] + B_n(c\cdot[\eta]^{\frac{3.4}{a}-1}\right]^{-1/n}\cdot\lambda_0^{-1} \tag{39}$$

The molar mass of the degraded sample can easily be calculated by assuming that the overlap parameter is equal to the intersection point of the horizontal with the line of slope 1.

$$\eta_{sp}(\dot{\gamma}) = (c\cdot[\eta])^*_{deg} \tag{40}$$

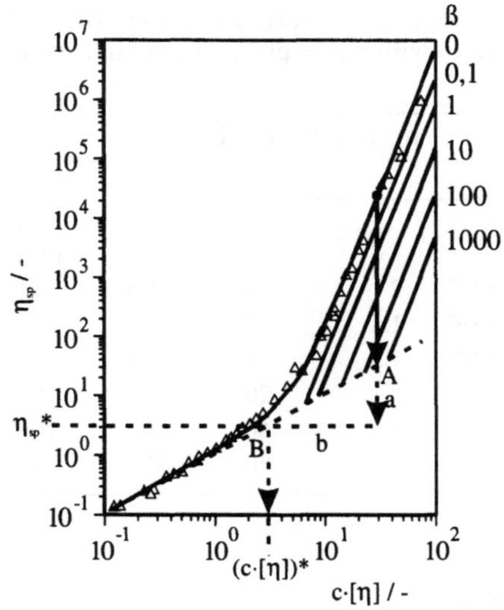

Fig. 18. Schematic plot of η_{sp} vs. the overlap parameter for polystyrene in toluene. The straight line with the slope 1 (η_{sp}=c [η]) divides the stable zone from the unstable

$$M_w = \left[\frac{\eta_{sp}(\dot{\gamma})}{c \cdot K_\eta}\right]^{1/a} \tag{41}$$

The correctness of this model has been shown viscometrically, as well as by means of wide-angle light scattering. Slope 1 illustrates that the diagram can be divided into two fields (Fig. 18). Below slope 1 there is no evidence of entanglements. In contrast to the traditional entanglement concept the slope was extended beyond $(c\cdot[\eta])^*$-values. At zero-shear conditions there is a maximum number of entanglements. By increasing $\dot{\gamma}$ the number of entanglements (in the power-law region) is reduced and if slope 1 is attained then degradation occurs, which is what this model predicts. Figure 18 demonstrates this precise tendency for polystyrene in toluene. The critical specific viscosity of a sheared polymer solution is the point of intersection (A) between the vertical (a) and the line with a slope of 1 (theoretical viscosity of a non-entangled fluid). Higher shear-stress leads to degradation. The mean molar mass of a degraded compound can easily be estimated by the point of intersection (B) between the horizontal (b) and the line with the slope of 1. The effect of shear rate dependent viscosity is calculated quantitatively using a modified reduced shear rate β':

$$\beta' = (\eta_0 - \eta_s)\cdot M^a \cdot \gamma / c^{1/a} \cdot R \cdot T \tag{42}$$

Normally β′ is used without considering the solvent quality. However, a master low curve can only be established using such a method. In Table 5 a comparison of the model prediction with the experimental findings, obtained by laser light scattering measurements, is given.

Table 5. Results from shear degradation measurements in two different rheometers (capillary, Couette-type) compared with model predictions of narrowly distributed polystyrene samples in toluene

$M_w \cdot 10^{-6}$ (g/mol)	c (wt%)	$\dot{\gamma}(s^{-1})$	η_{sp} (Pa·s)	$c\cdot[\eta]$	$c\cdot[\eta]^*$	$M_w \cdot 10^{-6}$ (g/mol)	
						Model prediction	Experimental findings
23.6	0.2	$1.5\cdot10^6$	2	4	1.5	6.0	6.4
9.2	0.2	$1.9\cdot10^6$	1.4	2.1	1	3.4	3.0
3.2	0.2	$2.3\cdot10^6$	1.2	1	0.9	2.8	2.8
23.6	1	5240	7	20	8	6.5	15.2
23.6	3	5240	42	60	48	16.7	16.6
23.6	4	5240	58	80	60	15.3	14.7

4
Elasticity in Shear Flow

When polymers are employed industrially, the elastic properties are just as significant as the viscous properties. The elastic material function may lead, under high shear rates (e.g. injection), to a loss of energy and to material defects. The complex shear modulus is composed of the storage modulus G′, which represents the elastic character, and the loss modulus G″, which represents the viscous behaviour. The loss factor $\tan\delta = G''/G'$ indicates whether a body is mainly elastic or viscous. For an ideal Newtonian fluid, the ratio is infinite for an ideally viscous and zero for an ideally elastic sample. It is known, however, that polymer fluids develop not only shear stresses but also normal stresses perpendicular to the flow directions. Normal stresses are primarily a manifestation of the elasticity of polymeric liquids and they are not found for Newtonian liquids. At high frequencies, the solutions show an increasing elastic portion, i.e. only a small part of the energy is transformed into viscous flow (small tanδ). Together with this linear viscoelastic behaviour, polymer solutions also exhibit, in contrast to Newtonian fluids, non-linear viscoelastic properties. A complete rheological characterization of polymer solutions and melts in steady shear flow is only given if the three stress functions (shear stress, τ_{12}, first and second normal stress differences, N_1 and N_2) are determined. In practice, no problems are usually encountered in measuring τ_{12} and N_1, but many difficulties occur when determining N_2. Additionally, there are some problems with flow irregularities in different kinds of steady shear flow at higher shear rates, which may be connected with general instabilities of coating processes, fiber spinning and calendaring [84]. These problems increase with increasing shear rate. These anomalous flows seem to be caused by the elastic component of the fluids and may be responsible for flow problems in polymer processing.

The precision of the measured N_1 values can be influenced by inertia. All N_1-values of shear rates higher than $10\ s^{-1}$ can be corrected for inertia effect using Eq. (43):

$$(N_1)_i = -\frac{3\rho\omega^2 R^2}{20}; \; N_1 = N_{1,measured} - (N_1)_i \qquad (43)$$

Inertia reduces the measured normal forces and it only depends on the diameter of the system and the rotational speed for a given solution. In the case of a polymer solution with real but small normal forces (dilute or lower molar masses), even negative N_1-values are simulated by this effect [85].

The N_2-values can be evaluated with the mathematical manipulations given by Petersen et al. [86] and developed by Jackson and Kaye [87].

$$N_2 = B \cdot [N_1 - \frac{2 \cdot F_{KPA}}{\pi \cdot (R')^2}]_{\dot{\gamma}_R = const.} - [\frac{F_{KPA} \cdot \partial \log F_{KPA}}{\pi \cdot (R')^2 \cdot \partial \log \dot{\gamma}_R}]_{\omega = const.} \quad (Pa) \qquad (44)$$

$$B = \frac{h + R' \cdot \tan\beta}{h}; \; \dot{\gamma}_R = \frac{\omega \cdot R'}{h + R' \cdot \tan\beta} \quad (s^{-1}) \qquad (45)$$

Considering the influence of shear rate on the viscoelastic behaviour of a non-Newtonian fluid, represented here by solutions of polystyrene standards in toluene, the obtained dependencies are plotted in Fig. 19. It can be seen that the shear stress is independent of the molar mass above a shear rate of 500 s^{-1} in the case of the molar masses chosen here. This is of fundamental importance in practice and theory (entanglement concept). The measured N_1 function shows much higher values than shear stress over almost all of the detectable shear range and a significant influence on the molar mass. In Fig. 19a the pure stresses are plotted, and the point of intersection, where the elastic force reaches a value as high as the viscous force, will be found in the region of relative low shear rate. At shear rates three decades higher, N_1 reaches values approximately one decade greater than the shear stress. Figure 19b shows the concentration dependence. In this case, a bigger difference of the viscous component in the power-law region yields. But in contrast, N_1 shows qualitatively the same behaviour in relation to the shear stress as in the molar mass dependence.

Deviation from laminar shear flow [88, 89], by calculating the material functions $\eta=f(\dot{\gamma})$, $\tau_{12}=f(\dot{\gamma})$, $\tau_{11}-\tau_{22}=f(\dot{\gamma})$, is assumed to be of a laminar type and this assumption is applied to Newtonian as well as viscoelastic fluids. Deviations from laminar flow conditions are often described as turbulent, as flow irregularities or flow instabilities. However, deviation from laminar flow conditions in cone-and-plate geometries have been observed and analysed for Newtonian and viscoelastic liquids in numerous investigations [90–95]. Theories have been derived for predicting the onset of the deviation of laminar flow between a cone and plate for Newtonian liquids [91–93] and in experiments reasonable agreements were found [95].

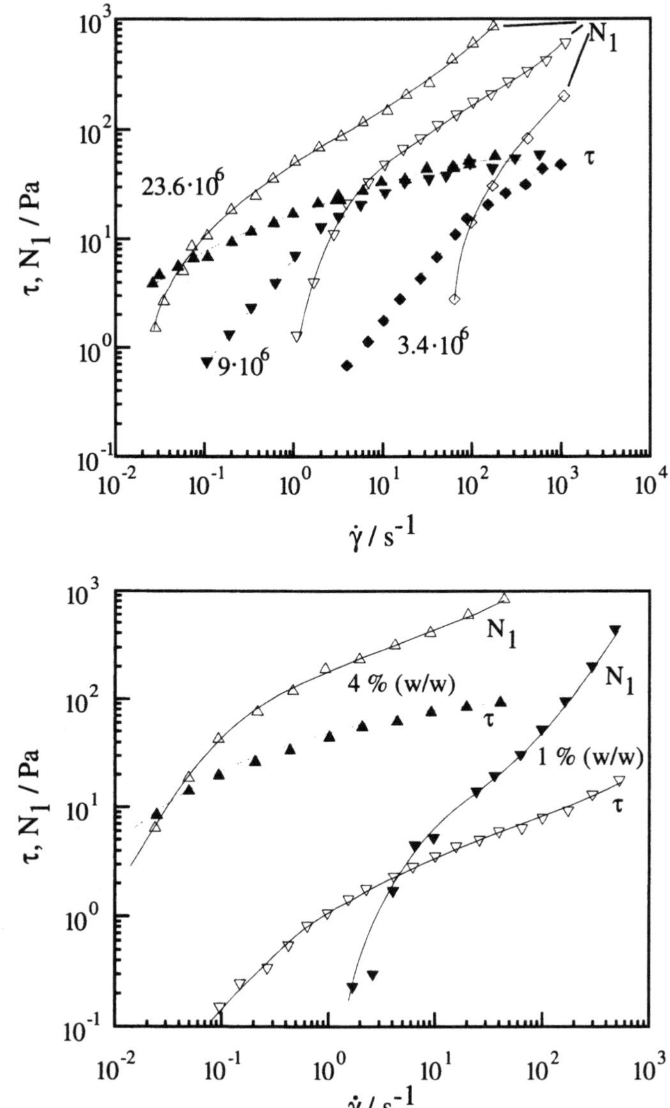

Fig. 19. Shear stress and first normal stress difference plotted as **a** function of shear rate for different molar masses, and **b** at different concentrations of polystyrene in toluene

Flow irregularities at gap angles of 30° were observed in viscoelastic liquids [94]. It has been indicated in theoretical treatments that the possibility of secondary flows [96, 97] in rotational devices is to be expected if the gap angle is much greater than 5°. For viscoelastic fluids deviations from laminar flow have only been reported in cone-and-plate geometries with gap angles above 10°.

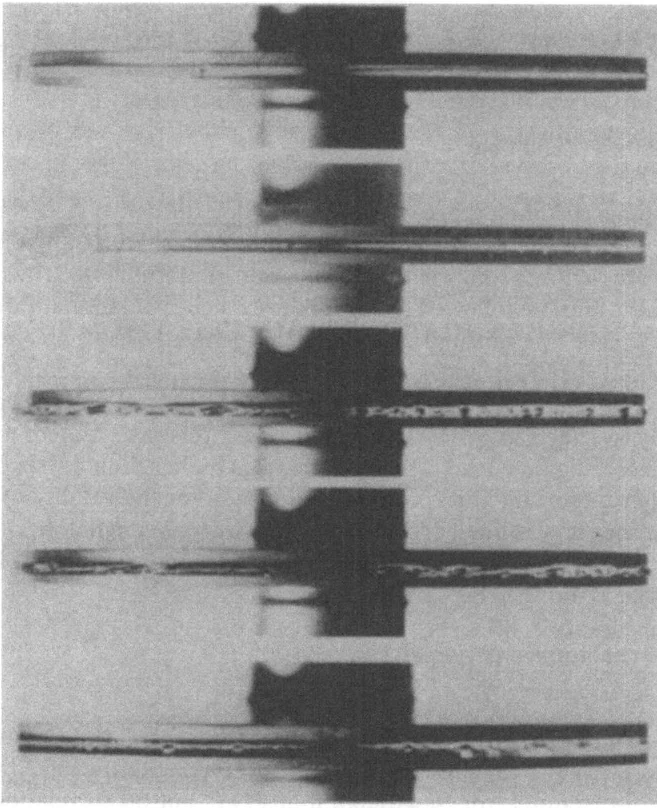

Fig. 20. Photographs of a secondary flow (vortices) in an aqueous PAAm solution at four shear rates in a cone-and-plate system. The cone is at the top. $M_\eta=2.53\cdot10^6$, c=4 wt%, T=23 °C, radius=5 cm, cone angle=0.04 rad (2.3°). **a** g=1.58 s^{-1}, **b** g=25 s^{-1}, **c** g=158 s^{-1}, **d** g=625 s^{-1}, **e** g=25 s^{-1}, enlargement of **b**

Flow irregularities were observed visually in steady shear at the edge of conventional cone-and-plate geometry with a cone angle of 2.3°. Figure 20a–e shows photographs of the flow irregularities in a cone and plate for an aqueous polyacrylamide (PAAm) solution. At low shear rates, the visible air surface is smooth but, on increasing the shear rate, flow irregularities become more pronounced. Figure 20e is an enlargement of Fig. 20b. The flow irregularities appear to be vortices.

Up to 100 vortices can be observed at one time, dependent on molar mass, concentration and shear rate. The vortices are seen to rotate about their own axis at the same time moving in the direction of shear. Vortices were seen up to the highest achievable shear rate g=1000 s^{-1}. Decreasing or increasing the shear rate has little or no influence on the generation of vortices. They appear or disappear in a few seconds. The vortices are not caused by air or vapour bubbles. Vortices were also observed in a solution to which a surfactant had been added, and the onset was at the same conditions as in an identical solution without surfactant. This demonstrates that a vortex formation is not a surface-induced phenomenon.

4.1
Detection of Onset Conditions

Cone-and-plate rheometers are widely used to determine the viscoelastic properties of polymer systems. The common assumption is that the flow is laminar, which provides a direct solution for the equation of motion [98]. Thus a number of precautions must be followed to ensure the existence of laminar flow between a cone and a plate. The gap angle should be low, generally less than about 4°. The most commonly used gap angles are indeed between 1 and 4°. Adams and Lodge [99] have shown that both the error in computed shear rate and the variation of shear rate across the gap will be less than 0.5% for this case. A theoretical explanation anticipating flow irregularities for angles higher than 4° has been given [96]. Another important and common assumption for the validity of the equation for laminar flow is that edge and inertial effects are negligible.

The transition from laminar to the onset of a secondary flow can be defined by the Reynolds number, Re:

$$\mathrm{Re} = \frac{\text{inertial force}}{\text{viscous force}} = \frac{\rho}{\eta} d^2 \omega \beta \tag{46}$$

Using this equation an attempt was made to find a critical Re-number which could be correlated to the onset of vortices observed with the naked eye, as has been done, for example, for Newtonian fluids [93], but no correlation could be found [88]. The reason is probably due to the fact that polymer solutions are viscoelastic fluids, also called second-order fluids.

In a first-order fluid (Newtonian) only significant dimensionless groups can be derived which include elastic behaviour [88].

$$S_R = \frac{\text{elastic force}}{\text{viscous force}} = \frac{\tau_{11} - \tau_{22}}{2 \cdot \tau_{12}} \tag{47}$$

The usual way is to plot S_R vs. shear rate.

In the case of polystyrene solutions sample loss was observed without pronounced overshoot taking place beforehand [88, 89] (Fig. 21). The intersection point gives conditions for the onset of flow irregularities. This method can also be applied when determining melt fractures which occur during extrusion [50]. Therefore one can conclude that by shearing a polymer solution, the expected smooth surface will be observed and, by increasing $\dot{\gamma}$, vortices may occur. In some cases, by further increasing $\dot{\gamma}$ pronounced overshoots are observed which may be followed by sample loss of the solutions. The existence of two kinds of flow irregularities was also found in capillary flow [100]. By using dimensionless analysis, it is possible to detect the onset of flow irregularities by plotting $S_R = f(\tau_{12})$ [88, 89].

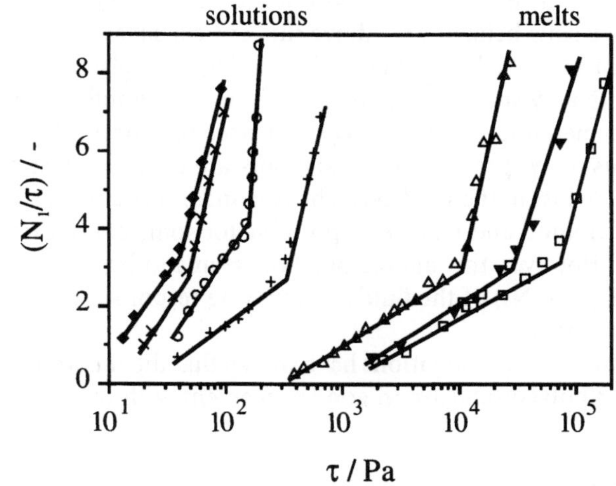

Fig. 21. Recoverable strain as a function of shear stress for polystyrene/toluene solutions with different molar masses at 5 wt%. (+) polyacrylamide/water solutions, (◆) 2 wt%, (x) 2 wt% (with surfactant), (O) 4 wt% and polyethylene melts

5
Conclusions

Viscoelastic properties have been discussed in relation to molar mass, concentration, solvent quality and shear rate. Considering the molecular models presented here, it is possible to describe the flow characteristics of dilute and semi-dilute solutions, as well as in simple shear flow, independent of the molar mass, concentration and thermodynamic quality of the solvent. The derivations can be extended to finite shear, i.e. it is possible to evaluate η as a function of the shear rate. Furthermore it is now possible to approximate the critical conditions (critical shear rate, critical rate of elongation) at which the onset of mechanical degradation occurs. With these findings it is therefore possible to tune the flow features of a polymeric solution so that it exhibits the desired behaviour under the respective deposit conditions.

Below a critical concentration, c^{**}, in a thermodynamically good solvent, η_0 can be standardised against the overlap parameter $c\cdot[\eta]$. However, for $c>c^{**}$, and in the case of a θ-solvent for parameter $c\cdot[\eta]\geq 0.7$, η_0 is a function of the Bueche parameter, cM_w. The critical concentration c^{**} is found to be M_w and solvent independent, as predicted by Graessley. In the case of semi-dilute polymer solutions the relaxation time and slope in the linear region of the flow are found to be strongly influenced by the nature of polymer–solvent interactions. Taking this into account, it is possible to predict the shear viscosity and the critical shear rate at which shear-induced degradation occurs as a function of M_w, c and the solvent power.

An increase in polymer molar mass and/or concentration and a decrease in the thermodynamic quality of the solvent will lead to a decrease in the critical shear rate at which the normal stress (elastic behaviour) will overwhelm the shear stress (viscous behaviour). Above this critical shear rate, the flow behaviour of the solutions will be strongly influenced by their elasticity. Therefore one can conclude that by shearing a polymer solution, a smooth surface will be observed as expected and by increasing $\dot{\gamma}$ vortices may occur. In some cases, by further increasing $\dot{\gamma}$, pronounced overshoots are observed which may be followed by sample loss of the solutions. The reason for this unusual flow behaviour is not yet understood and may be responsible for some contradictory results in the literature. However, this anomalous flow seems to be caused by the elastic component (N_1 and N_2) of the fluids and may be responsible for flow problems in polymer processing.

A great number of investigations have shown that the theoretical correlations established for polystyrene are in good agreement with the experimental data [83, 101–107].

Acknowledgements. The authors would like to thank the European Commission (Direktorate General XII: Science, Research and Development; Direktorate G: Human Capital and Mobility) for their financial support.

The authors would also like to thank the heads of the research groups and their co-workers, namely P. Navard (Armines, France), G. Marruci (Naples, Italy), M. Cates (Cambridge, England), and B. Simionescu (Jassy, Rumania) for fruitful and stimulating discussions.

References

1. Cheftel JC, Cuq JL, Lorient D (1992) Lebensmittel-Proteine. Behr's, Hamburg, p 88
2. Harris P (1990) Food gels. Elsevier Applied Science, New York, p 208
3. Walter RH (1998) Polysaccharide association structures in food. Marcel Dekker, New York, p 289
4. Lapasin R, Pricl S (1995) Rheology of industrial polysaccharides. Theory and applications. Padstow, Cornell, p 135
5. Laba D (1993) Rheological properties of cosmetics and toiletries. Marcel Dekker, New York, p 55
6. Casale A, Porter RSA (1978) Polymer stress reactions, vol 1. Academic Press, New York, p 70
7. Koedritz LF, Harvey AH, Honarpour M (1989) Introduction to petroleum reservoir analysis. Gulf, Houston
8. Gampert B (1985) The influence of polymer additives on velocity and temperature fields. Springer, Berlin Heidelberg New York, p 371
9. Kulicke W-M, Gräger H, Kötter M (1989) Adv Polym Sci 89:1
10. Glass JE. (1986) Water-soluble polymers. Beauty with performance. American Chemical Society, Washington, DC, p 183
11. Hebeish A, El-Zairy MR, El-Rafie MH, Higazy A, El-Sisy F (1991) Starch 43:98
12. Woffindin C, Hoenich NA (1992) JNS Nephrol Dialysis Transplant 7:340
13. Beretka J (1992) J Chem Technol BioTechnol 55:269
14. Schulz DN, Glass JE (1991) Polymers as rheology modifiers. American Chemical Society, Washington, DC, p 322
15. Carr ME (1992) Starch 44:219
16. Paine AJ (1990) J Colloid Interface Sci 138:157
17. Cohen E (1993) Arch Insect Biochem Physiol 22:245
18. Glass JE, Swift G (1990) Agricultural and synthetic polymers. American Chemical Society, Washington, DC, p 33

19. Kniewske R, Kulicke W-M (1983) Makromol Chem 184:2173
20. Kulicke W-M, Kniewske R (1984) Rheol Acta 23:75
21. Kulicke W-M, Griebel Th, Bouldin M (1991) Polymer News 16:39
22. Bouldin M, Kulicke W-M, Kehler H (1988) Colloid Polym Sci 266:793
23. Kulicke W-M (1986) Fließverhalten von Stoffen und Stoffengemischen. Hüthig & Wepf, Basel, p 218
24. Kulicke W-M, Klein J, Kniewske R (1982) Progress Polym Sci 8:373
25. Kulicke W-M, Haas R (1984) Ind Eng Chem Fundam 23:308
26. Haas R, Kulicke W-M (1984) Ind Eng Chem Fundam 23:316
27. Schramm G (1994) A practical approach to rheology and rheometry. Haake, Karlsruhe, p 146
28. Carreau PJ, De Kee DCR, Chhabra RP (1997) Rheology of polymeric systems. Principles and applications. Hanser, Munich, p 35
29. Macosko CW (1994) Rheology. Principles, measurements and applications. VCH, New York, p 86
30. Bird RB, Armstrong RC, Hassager O (1977) Dynamics of polymeric liquids. Wiley, New York
31. Mark JE, Eisenberg A, Graessley WW, Mandelkern L, Koenig JL (1984) Physical properties of polymers. American Chemical Society, Washington, DC
32. Barnes HA, Hutton JF, Walters K (1989) An introduction to rheology. Elsevier, Amsterdam
33. Rouse PE (1953) J Chem Phys 21:1272
34. Zimm BH (1956) J Chem Phys 24:269
35. Bueche F (1952) J Chem Phys 20:1959
36. Debye P, Bueche F (1948) J Chem Phys 16:573
37. Bueche F (1956) J Chem Phys 25:599
38. Berry GC, Fox TG (1968) Adv Polym Sci 5:261
39. Ferry JD, Landel RF, Williams ML (1955) J Appl Phys 26:359
40. Graessley W, Hazelton R, Lindeman R (1967) Trans Soc Rheol 11:267
41. Graessley W (1967) J Chem Phys 47:1942
42. Graessley W (1965) J Chem Phys 43:2696
43. De Gennes PG (1971) J Chem Phys 55:572
44. De Gennes PG (1979) Scaling concepts in polymer physics. Cornell University Press, Ithaca, New York
45. Leger L, De Gennes PG (1982) Annu Rev Phys Chem 33:49
46. Doi M (1983) J Polym Sci Polym Phys Ed 21:667
47. Doi M, Edwards SF (1986) The theory of polymer dynamics. Oxford University Press, p 234
48. Doi M, Edwards SF (1978) J Chem Soc Farad Trans 74:1802
49. Han CD, Jhon MS (1986) J Appl Polym Sci 32:3809
50. Han CD (1976) Rheology in polymer processing. Academic Press, London, p 61
51. Forsman WC (1989) Polymers in solution. Theoretical considerations and newer methods of characterization. Plenum Press, New York, p 145
52. Lodge AS (1964) Elastic liquids. Academic Press, New York
53. Simha R, Zakin L (1962) J Coll Sci 17:270
54. De Gennes PG (1976) Macromolecules 9:587
55. Klein J (1978) Macromolecules 11:852
56. Fujita H (1990) Polymer solutions. Elsevier, Amsterdam, p 182
57. Graessley WW (1974) Adv Polym Sci 16:49
58. Baumgärtel M, Willenbacher N (1996) Rheol Acta 35:168
59. Masuda T, Kitagawa K, Onogi S (1970) Polymer J (Japan) 1:418
60. Kulicke W-M, Klare J (1980) Angew Makromol Chem 84:67
61. Casale A, Moroni A, Civardi E (1976) Angew Makromol Chem 53:1
62. Casale A, Porter RS, Johnson JF (1971) J Macromol Sci-Rvs Macromol Chem C5:387
63. Huggins ML (1942) J Am Chem Soc 64:2716

64. Daoud M, Cotton JP, Farnoux B, Jannink G, Sarma G, Benoit H, Dupressix R, Picot C, De Gennes PG (1975) Macromolecules 6:804
65. Zakin JL, Wu R, Luh H, Mayhan KG (1976) J Polym Sci Polym Phys Ed 14:299
66. Schurz J (1975) Rheol Acta 14:293
67. Onogi S, Kobayashi TKojima Y, Taniguchi Y (1963) J Appl Polym Sci 7:847
68. Griebel TH, Kulicke W-M, Kniewske R (1992) Mehl und Brot 5:154
69. Arendt O, Kulicke W-M (1998) Angew Makromol Chem 259:61
70. Böhm N., Kulicke W-M (1999) Carbohydr Res 293
71. Robinson G, Ross-Murphy SB, Morris ER (1982) Carbohydr Res 107:17
72. Abdel-Alim AH, Balke ST, Hamielec AE (1973) J Appl Polym Sci 17:1431
73. Attane P, LeRoy P, Picard JM, Turrel GJ (1981) Non-Newtonian Fluid Mech 9:13
74. Ferry JD (1978) Pure Appl Chem 50:299
75. Ferry JD (1980) Viscoelastic properties of polymers, 3rd edn. Wiley, New York, p 38
76. Stratton RA (1966) J Colloid Interface Sci 22:517
77. Wissbrunn KF, Metzner AB, Rangel-Nafaille C (1984) Macromolecules 17:1187
78. Kehler H, Kulicke W-M (1986) Chem Eng Technol 10:802
79. Elias HG (1996) Polymere von Monomeren und Makromolekülen zu Werkstoffen eine Einführung. Hüthig & Wepf, Heidelberg, p 273
80. Vinogradov GV, Malkin AV (1980) Rheology of polymers. Springer, Berlin Heidelberg New York, p 128
81. Williams MC (1967) AIChE J 13:534
82. Cross MM (1969) J Appl Polym Sci 13:765
83. Kulicke W-M, Porter RS (1981) J Polym Sci Ed 19:1173
84. Middleman S (1977) Fundamentals of polymer processing. McGraw-Hill, New York, chap 15
85. Kulicke W-M, Kiss G, Porter RS (1977) Rheol Acta 16:568
86. Petersen JF, Rautenbach R, Schümmer P (1975) Rheol Acta 14:968
87. Jackson R, Kaye A (1966) Br J Appl Phys 17:1355
88. Kulicke W-M, Porter RS (1979) J Appl Polym Sci 23:953
89. Kulicke W-M, Jeberien HE, Kiss G, Porter RS (1979) Rheol Acta 18:711
90. Pearsom JRA (1976) J Fluid Mech 4:163
91. Walters K (1975) Rheometry. Wiley, New York, p 65
92. King MJ, Waters ND (1970) Rheol Acta 9:164
93. Fewell ME, Hellums JD (1974) AIChE Winter Conf Washington, DC, Dec 1–5
94. Turian RM (1972) Ind Eng Chem Fundam 11:361
95. Gleisle W (1976) Rheol Acta 15:305
96. Giesekus H (1965) Rheol Acta 4:85
97. Lodge AS (1964) Elastic liquids. Academic Press, New York
98. Lodge AS (1974) Body tensor fields in continuum mechanics. Academic Press, New York
99. Adams N, Lodge AS (1964) Phil Trans R Soc London A256:149
100. Southern JH, Paul DR (1974) Polym Eng Sci 14:560
101. Kulicke W-M, Böse N (1982) Polym Bull 7:205
102. Kulicke W-M, Hörl H-H (1983) Angew Makromol Chem 116:149
103. Oertel R, Kulicke W-M (1991) Rheol Acta 30:140
104. Kulicke W-M, Reinhardt U (1993) Polym Mat Sci Eng 69:491
105. Reinhardt UT, Eidam D, Kulicke W-M (1994) J Getreide, Mehl und Brot 4:56
106. Kulicke W-M, Duhm L, Schuch A (1994) Chem-Ing-Technik 12:1643
107. Kulicke W-M, Kull AH, Kull W, Thielking H, Engelhardt J, Pannek J-B (1996) Polymers 13:2723

Received: March 1999

Bridging the Gap Between Atomistic and Coarse-Grained Models of Polymers: Status and Perspectives

Jörg Baschnagel[1,6], Kurt Binder[1], Pemra Doruker[2], Andrei A. Gusev[3], Oliver Hahn[4], Kurt Kremer[4], Wayne L. Mattice[2], Florian Müller-Plathe[4], Michael Murat[4,5], Wolfgang Paul[1], Serge Santos[3], Ulrich W. Suter[3], Volker Tries[1]

[1] Institut fur Physik, Johannes-Gutenberg Universität Mainz, D-55099 Mainz, Staudinger Weg 7, Germany
[2] Institute of Polymer Science, The University of Akron, Akron, Ohio 44325-3909, USA
[3] Department of Materials, Institute of Polymers, ETH, CH-8092 Zürich, Switzerland
[4] Max-Planck-Institut fur Polymerforschung, D-55021 Mainz, Germany
[5] Soreq Nuclear Research Center, Yavne 81800, Israel
[6] Institut Charles Sadron, 6rue Boussingault, 67083 Strasbourg, France

Recent developments that increase the time and distance scales accessible in the simulations of specific polymers are reviewed. Several different techniques are similar in that they replace a model expressed in fully atomistic detail with a coarse-grained model of the same polymer, *atomistic → coarse-grained (and beyond!)*, thereby increasing the time and distance scales accessible within the expenditure of reasonable computational resources. The bridge represented by the right-pointing arrow can be constructed via different procedures, which are reviewed here. The review also considers the status of methods which reverse this arrow, *atomistic ← coarse-grained*. This "reverse-mapping" recovers a model expressed in fully atomistic detail from an arbitrarily chosen replica generated during the simulation of the coarse-grained system. Taken in conjunction with the efficiency of the simulation when the system is in its coarse-grained representation, the overall process

$$\begin{array}{ccc} & \text{I. Atomistic, t(0)} & \rightarrow & \text{II. Coarse - grained, t(0)} \\ & & & \downarrow \\ \text{abc} & \text{IV. Atomistic, t(0) + t} & \leftarrow & \text{III. Coarse - grained, t(0) + t} \\ & & & \downarrow \\ & & & \text{V. Semimacroscopic} \end{array}$$

permits a much more complete equilibration of the system (larger effective size of Δt) when that equilibration is performed with the coarse-grained replicas (II → III) than if it were attempted with the fully atomistic replicas (I → IV).

Keywords. Atomistic models, Bridging, Coarse-grained models, Polymers, Simulations

Advances in Polymer Science, Vol. 152

1 Introduction

All the macroscopic properties of polymers depend on a number of different factors; prominent among them are the chemical structures as well as the arrangement of the macromolecules in a dense packing [1–6]. The relationships between the microscopic details and the macroscopic properties are the topics of interest here. In principle, computer simulation is a universal tool for deriving the macroscopic properties of materials from the microscopic input [7–14]. Starting from the chemical structure, quantum mechanical methods and spectroscopic information yield effective potentials that are used in Monte Carlo (MC) and molecular dynamics (MD) simulations in order to study the structure and dynamics of these materials on the relevant length scales and time scales, and to characterize the resulting thermal and mechanical proper-

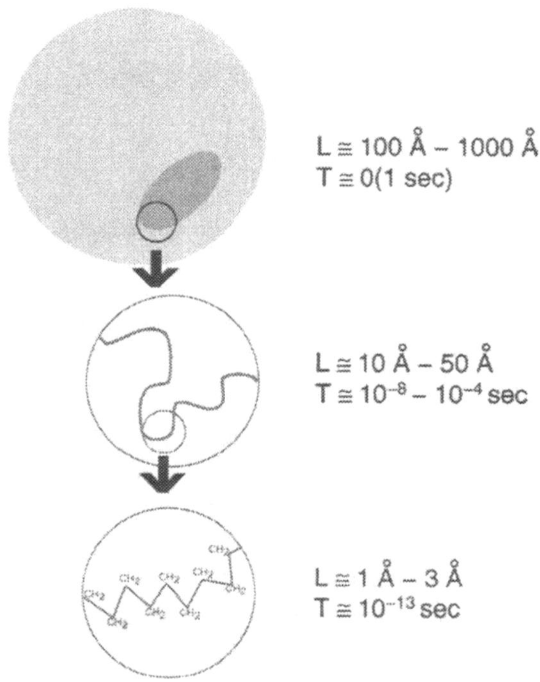

Fig. 1.1. Sketch of the different time and length scales in polymer problems. Starting from the top, a marked chain in a melt or dense solution of otherwise identical chains can be observed only as a very pale shadow. The typical extension of the shadow is given by the diameter of the overall coil, as indicated. The characteristic time for this picture to change can vary dramatically depending on chain length and temperature, starting at about 10^{-4} s for short chains and "high temperatures", with essentially no upper limit. Looking more closely, more of the chain structure is revealed. This is the universal entropy dominated coil regime. Again the variation in time is very large, especially as a function of temperature. Typical times, as they are present in typical experiments, are indicated. Only if the figure is examined much more closely can the chemical details of the polymers be identified. There the behavior is governed by the local chemical details of the species under consideration and is energy dominated. The lower time boundary is given by the highest frequency, which usually is from the covalent bond oscillations. Typical coarse-grained simulations are situated somewhere in between the coil and the microscopic regime. From [15]

ties. It is tempting to perform a computer simulation of a melt of polymers where all the details of the chemical structure of the monomers are included. Such a simulation should then provide complete information about the properties of the system under investigation. However, consideration of the interesting properties of polymers and of available computational resources strongly suggests that a different approach should be taken.

As for all disordered complex macromolecular materials, polymers can be characterized by a hierarchy of different length and time scales, and these scales span an extremely wide range, as outlined in Fig. 1.1 [15]. The diffusion

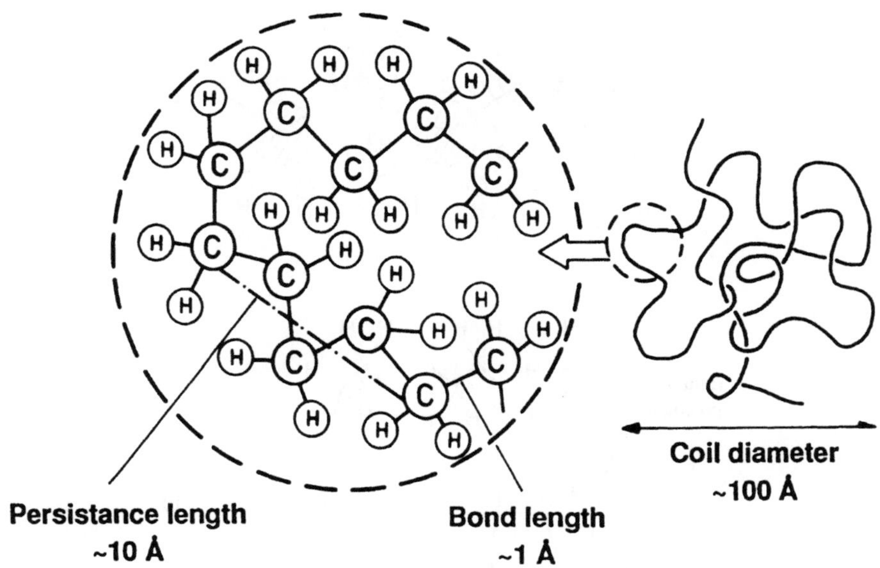

Fig. 1.2. Schematic illustration of the various length scales of a macromolecule. Polyethylene is used as an example. From [17]

constant D could be measured by monitoring the mean square displacement of the center of gravity of the chains. For polymer melts, however, the relevant length and time scales are a serious problem for any simulation [12, 15–20]. It is almost trivial to show that even with the present high-speed computers such a simulation is simply impossible. Even for the biggest computers it would need ages of computer time, and certainly exceed the life time of any scientist. On the other hand, even if this were possible, it is very questionable whether such an attempt should even be undertaken. Such a simulation would provide an enormous amount of data with almost all the generated information actually irrelevant to the questions under consideration. In order to understand the system and to be able to make suggestions for material improvement it is important to structure the information and understand the results, it is not sufficient to just report data.

A polymer coil does not only possess a structure on the atomistic scale of a few Å, corresponding to the length of covalent bonds and interatomic distances: characteristic of macromolecules are coils that more or less, obey Gaussian statistics and have a diameter of the order of hundreds of Å (Fig. 1.2) [17]. Structures of intermediate length scales also occur e. g., characterized by the persistence length. For a simulation of a polymer melt, one should consider a box that contains *many* such chains that interpenetrate each other, i. e., a box with a linear dimension of several hundred Å or more, in order to ensure that no artefacts occur attributable to the finite size of the simulation box or the periodic boundary conditions at the surfaces of the box. This ne-

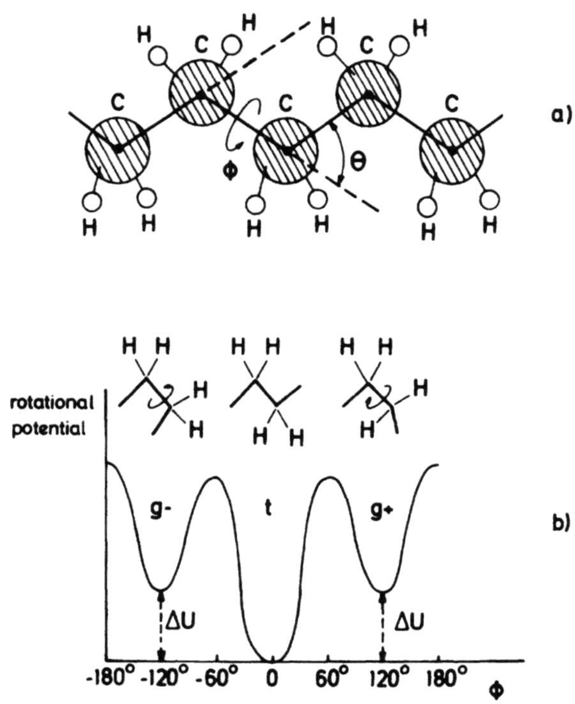

Fig. 1.3. **a** Schematic model of a piece of a polyethylene chain. The hydrogen atoms (H) are not treated explicitly in the "united atom" approximation, but rather denoted as effective spherical segments (shaded) representing a whole CH_2 unit. The segments are connected by harmonic bonds, shown as straight lines, with bond lengths ℓ. Segments are labeled consecutively by an index i, i = 0 to i = N_p−1. Three successive segments i−1, i, i+1 define a bond angle θ_i and four successive segments i−2, i−1, i, i+1 define a rotational angle ϕ_i, namely the angle between the plane spanned by the segments i−2, i−1, i and the plane formed from i−1, i, i+1. Note that all of the ϕ_i's are zero in the all-trans configuration drawn here. **b** Qualitative sketch of the torsional potential for *n*-alkane chains, indicating the three energetically preferred states "gauche minus" (g^-), "trans" (t), "gauche plus" (g^+). The minimum of the trans configuration is deeper by an amount ΔU. From [16]

cessity to simulate very large systems containing many millions of atoms is only one difficulty hampering the simulation of chemically realistic models of polymeric materials, however.

An even more serious problem concerns the corresponding time scales: on the most microscopic level, vibrations of bond lengths and bond angles have characteristic times of approx. $\tau_{vib} \approx 10^{-13}$ s; somewhat slower are the jumps over the barriers of the torsional potential (Fig. 1.3), which take place with a time constant of typically $\omega^{-1} \approx 10^{-11}$ s. On the semi-microscopic level, the time that a polymer coil needs to equilibrate its configuration is at least a factor of the order N_p^2 larger, where N_p is the degree of polymerization, $\tau = \omega^{-1} N_p^2$. This formula applies for the Rouse model [21,22], i. e., for non-entangled melts. If we consider chains with chain lengths N_p exceeding the entanglement chain length N_e, the coil relaxation time τ rather scales according to the reptation model [22,23], i. e., $\tau \approx \omega^{-1} N_p^3/N_e$ (prefactors of order unity are disregarded here). For polyethylene (PE) [24] $N_e \approx 100$ and thus this estimate for the relaxation time yields roughly $\tau \approx 10^{-7} - 10^{-4}$ s for a degree of polymerization in the range $10^2 \leq N_p \leq 10^3$. For very high temperatures the time τ may be somewhat smaller, but at lower temperatures the time τ may be much larger, up to macroscopic times near the glass transition temperature [25].

Many practically important polymers have a chemical structure that is considerably more complicated than PE, and this fact further complicates the simulation of macromolecular materials. As a consequence of all these arguments, it is clear that a simulation of fully atomistic models of a sufficiently large system over time scales for which thermal equilibration could be reached at practically relevant temperatures, is absolutely impossible; thus a different approach must be taken!

On the other hand, it is important to relate the chemical structure of a system to the overall behavior of a material [26]. Thus, one of the long standing challenges when modeling complex materials is to apply methods which allow an investigator to cover regimes in the range from microscopic to semi-macroscopic [27–32]. Any approach other than this (impossible) brute force approach must make compromises, as far as the completely realistic modelling of polymeric materials with all their details is concerned. Different groups tend to make rather different compromises, depending on what features of the problem they consider particularly important. The current status of several approaches on connecting scales on different levels is reviewed here.

The review commences, in Section 2, at the level of the electronic degrees of freedom (Müller-Plathe), which takes one from electrons to atoms, as well as an effective potential operating between atoms.

There is great interest in the development of methods that allow the identification of a reasonably "good" structure with which to start the simulation of dense atomistically detailed polymer systems. The problem of generating dense polymer systems is formidable due to the high density and the connectivity of polymer systems. For crystal structures this can be systematically achieved [33,34]; for amorphous structures, however, there is no generally satisfactory method available. Two recent developments in methods for generating amorphous packing (Santos, Suter) are reviewed in Section 3.

A suitable approach to the equilibration of an amorphous polymer system at bulk density becomes much more likely when the fully atomistic model in continuous space is replaced by an equivalent coarse-grained model on a lattice with sufficient conformational flexibility. Different strategies, which seek results at different levels of detail, can be employed to create an appropriate coarse-grained model. Section 4 (Doruker, Mattice) describes an approach which attempts to retain a connection with the covalent bonds in the polymer. The rotational isomeric state (RIS) [35,36] model for the chain is mapped into a coarse-grained model on a high-coordination lattice, and later the mapping is reversed to recover a fully atomistic description in continuous space. Section 5 (Binder, Paul, Tries, Baschnagel) describes a second approach, where there are additional advantages in time and distance scales, achieved with a slight additional sacrifice in detail at the Å level. It exploits the versatility of the bond fluctuation model [37–42], by mapping the chemically realistic model of a polymer to its equivalent chain.

Recently a hierarchy of methods has been developed which covers the mapping of polymers to a mesoscopic level as well as the reintroduction of the atomistic structure [43–45]. Section 6 (Kremer, Murat, Hahn) gives some very first attempts to bridge the gap from microscopic to mesoscopic [43,44] and thereafter to the semi-macroscopic regime [45] within a simulation scheme.

Application to heterogeneous polymer solids, and elastic composites, is presented in the Section 7 (Gusev, Suter), which is followed by a summary and the outlook for the various methods reviewed here. It will be apparent to the reader that this review thus assembles several building blocks for the difficult task to bridge the gaps from the atomistic to the macroscopic scales in space and times for the simulation of polymeric materials. Integrating these building blocks into one coherent framework still is not fully solved and a matter of current research.

2 Combining Classical and Quantum-Mechanical Approaches in Describing Polymer Systems

The most fundamental level at which polymeric systems are usefully described includes electronic degrees of freedom. These are directly responsible for a number of polymer properties: non-linear optical properties, electronic conductivity, polymerisation stereochemistry and kinetics, photochemistry, (photo)chemical degradation, to name but a few. These properties cannot be described or explained even qualitatively without considering the electronic wave function. On the other hand, there are properties derived from the molecular electron distribution which may be viewed after abstracting from the electrons, such as: the polarisability, molecular charge distribution and electrostatic potentials, dispersion (van der Waals) interactions and so on. Properties of the second class originate in the behavior of the electrons but they are usually treated in an effective way by implicitly averaging over the electronic degrees of freedom.

The usual way chemistry handles electrons is through a quantum-mechanical treatment in the frozen-nuclei approximation, often incorrectly referred to as the Born-Oppenheimer approximation. A description of the electrons involves either a wavefunction ("traditional" quantum chemistry) or an electron density representation (density functional theory, DFT). Relativistic quantum chemistry has remained a specialist field and in most calculations of practical interest for polymer research it is assumed that relativistic effects are either irrelevant or can be parameterized into suitable pseudopotentials.

It is the purpose of this section to review ways in which processes involving electrons are either explicitly accounted for in calculations on polymeric systems or in which a more or less rigorous abstraction from the electronic degrees of freedom into effective models of a coarser-grained nature is performed. The "next level up" from electrons is obviously atoms. Hence, this section deals mainly with the connection between quantum chemistry and atomistic (force field) simulations. Calculations which exclusively use quantum chemistry are not covered. This excludes, for example, all of the recent work on metallocene catalysis.

2.1
Quantum Chemistry Based Force Fields

The most straightforward way of using quantum-chemical information at the atomistic level is by means of a force field. This requires that no truly electronic processes (chemical reactions, electronically excited states) are involved and that basically a description of a (macro)molecule in its closed-shell ground state is desired.

An atomistically detailed force field typically contains at least terms for describing the following types of interactions:

(i) bonded interactions: chemical bonds, bond angles, dihedral angles, harmonic (improper) dihedral angles, etc. Some force fields sport cross terms between different bonded interactions.
(ii) nonbonded interactions: short-range repulsion (excluded volume interaction, Pauli repulsion), dispersion attraction (van der Waals interaction, London forces), electrostatic interactions (Coulomb forces). Some force fields include polarisability (induced dipoles), some have special functional forms to describe hydrogen bonding.

The usefulness of quantum-chemical methods varies considerably depending on what sort of force field parameter is to be calculated (for a detailed discussion, see [46]). There are relatively few molecular properties which quantum chemistry can provide in such a way that they can be used directly and profitably in the construction of a force field. Quantum chemistry does very well for molecular bond lengths and bond angles. Even semiempirical methods can do a good job for standard organic molecules. However, in many cases, these are known with sufficient accuracy: a C–C single bond is 1.53 Å except under exotic circumstances. Similarly, vibrational force constants can often be transferred from similar molecules and need not be recalculated.

In the other extreme, quantum chemistry is impractical for Lennard-Jones parameters of most molecules of interest to simulators, since the description of dispersive interactions needs huge basis sets and a high-order treatment of electron correlation. DFT methods seem to have similar difficulties, since in purely dispersive systems local density (LDA) methods give a strong overbinding, whereas gradient-corrected functionals lead to no attraction at all [47]. However, there seems to have been some progress in estimating dispersion coefficients by perturbative methods, i. e., from calculations on single molecules, rather than, supermolecules. The classical route is via the frequency-dependent polarisability and the Casimir-Polder equation [48]. Recently, these concepts have been adapted in the framework of DFT [49]. Atomic partial charges can be derived from wave functions. However, the fundamental problem (the atomic partial charge is not an observable and, hence, there is no well-defined quantum-mechanical operator for it) leads to technical problems (there are many recipes for extracting partial charges from a wave function, all of which give different answers) as well as language problems (the quantum chemist uses partial charges as a data reduction concept to approximately characterize the distribution of electrons in a molecule; the molecular simulator views them as effective force field parameters, i. e., as mere numbers). Notwithstanding, there seem to be recipes which for certain molecules allow the extraction of partial charges that can be directly used in a force field. In particular, fits of the charges to the molecular electrostatic potential are used. The latter method, may produce spurious results due to an effective underdetermination of the fitting problem [50]. It should also be kept in mind that atomic partial charges may be conformation dependent, particularly for extended π-systems [50A].

One additional important reason why nonbonded parameters from quantum chemistry cannot be used directly, even if they could be calculated accurately, is that they have to implicitly account for everything that has been neglected: three-body terms, polarization, etc. (One should add that this applies to experimental parameters as well: A set of parameters describing a water dimer in vacuum will, in general, not give the correct properties of bulk liquid water.) Hence, in practice, it is much more useful to tune these parameters to reproduce thermodynamic or dynamical properties of bulk systems (fluids, polymers, etc.) [51–53]. Recently, it has been shown, how the cumbersome trial-and-error procedure can be automated [54–56A].

On the bright side, there is an area where quantum chemistry is very successful, namely in the parameterization of dihedral angle (torsional) interactions. Except for tiny molecules, there is no way to determine torsional barriers or the energetic differences between conformations (say, trans and gauche) which is as quick and reliable as ab initio quantum chemistry. Often Hartree-Fock or density functional calculations in a polarized split-valence basis do a very nice job [57], whereas semiempirical methods should be used with extreme care [58]. It should be noted that the determination of molecular geometries and force field parameters from quantum-chemistry works as well (or as badly) for electronically excited states as it does for ground states, except for the usually higher computational cost. For excited state molecules, quantum chemistry is often the only way of estimating parameters, since parameterization against experiment is impossible.

A number of polymer force fields have been developed using quantum chemistry input, usually following the thoughts laid out above, i. e., using ab initio quantum chemistry to derive bonded parameters, whilst parameterizing the nonbonded part against experimental bulk properties. A force field for poly(tetrafluoroethylene) captured correctly the split in the trans and gauche minima [59]. All the recently developed force fields for polyethylene oxide [57,60–62] relied on high quality ab initio studies of oligomers. Another example is a new force field for poly(vinyl alcohol) [63] which combines quantum-chemical input for torsional terms with nonbonded interactions parameterized to reproduce liquid ethanol and ethanol/water mixtures [51].

2.2
Concepts from Quantum Chemistry

It should not be forgotten that quantum-chemical calculations can provide physical and chemical understanding in addition to hard numbers. Often, such an insight obtained from an electronic structure calculation leads to a useful concept or approximation in subsequent molecular simulation or analytical model building.

An example may serve to illustrate the idea. Conformational equilibria of poly(ethylene oxide) and its dimer, the dimethoxyethane molecule (ethylene glycol dimethyl ether, glyme), have puzzled researchers for some time. This molecule has three rotatable dihedral angles involving heavy atoms only, which can assume trans, gauche$^+$ or gauche$^-$ conformations, giving rise to ten different molecular conformers. Of these, the all-trans (ttt), trans-gauche-trans (tgt) and the trans-gauche$^+$-gauche$^-$ (tgg$^-$) have the lowest energies. While all quantum-chemical calculations in vacuum have unanimously found ttt to be the minimum, all experimental measurements in condensed phases (crystal, melt, solution) found mainly tgt. This effect has nothing to do with the level of the quantum-chemical treatment (as was believed for a long time), but is due to the different environment (vacuum vs. condensed phase). It could be reproduced by a simple self-consistent reaction-field treatment [64] which just assumed a polarizable dielectric continuum around the glyme molecule. The shift in relative stability is caused by the large difference in the dipole moments of ttt (zero by symmetry) and tgt (1.4 D). The dipole moment causes polarization of the continuum and, consequently, stabilization of the polar conformation. Although the energy by which the tgt conformer is stabilized in a reaction-field treatment (2.3 kJ/mol) does not agree quantitatively with what is estimated from spectra and poly(ethylene oxide) (PEO) chain conformations (more like 4–5 kJ/mol), the simple treatment shows that a polarizable environment produces a physical effect of the right order of magnitude and in the right direction. This convinced the authors that a successful PEO force field has to allow for the dielectric surrounding of the polymer. They proceeded to develop a force field that, in vacuum, produces ttt conformers and, in solution, tgt conformers, as appropriate. In designing the force field, however, the experimental stabilization energy, rather than, its quantum chemical counterpart, was used [60, 65].

2.3 Hybrid Simulations

Very often, a system can be conceptually partitioned into an "interesting" primary system (reactants, products, excited-state molecule, catalytic center, enzyme active site, etc.) and an environment (solvent, surface, protein, polymer matrix, etc.) which is not in itself interesting but which is known or assumed to have an influence on what goes on in the primary system. Moreover, the primary system is often small (one molecule, a few molecules or fragments of molecules) whereas the environment is huge. For many problems the split is like a few dozen atoms for the primary system versus tens of thousands of atoms for the environment. Describing the tens of thousands "uninteresting" environment degrees of freedom quantum-chemically would not only be a waste of resources but would make the calculation completely unfeasible. However, the conceptual partitioning of the system outlined above has been translated into hierarchical computational schemes: An accurate and expensive quantum-chemical treatment of the interesting small primary part of the system and a cheap force field for the large and uninteresting environment is employed. Such schemes are known as hybrid schemes or quantum-mechanical/molecular-mechanical (QM/MM) methods [65A, 65B].

The QM/MM Hamiltonian is split into the three contributions

$$H = H_{pp} + H_{ee} + H_{pe} \tag{2.1}$$

The forces on the atoms are contributed by all three parts. H_{pp} acts on the primary atoms, H_{ee} on the environment atoms and H_{pe} on both. With all forces evaluated, all atoms are propagated classically to their next position using the chosen molecular dynamics, stochastic dynamics, or Monte Carlo scheme.

H_{pp} describes the primary system by a quantum-chemical method. The choice is dictated by the system size and the purpose of the calculation. Two approaches of using a finite computer budget are found: If an expensive ab-initio or density functional method is used the number of configurations that can be afforded is limited. Hence, the computationally intensive Hamiltonians are mostly used in geometry optimization (molecular mechanics) problems (see, e. g., [66]). The second approach is to use cheaper and less accurate semi-empirical methods. This is the only choice when many conformations are to be evaluated, i. e., when molecular dynamics or Monte Carlo calculations with meaningful statistical sampling are to be performed. The drawback of semi-empirical methods is that they may be inaccurate to the extent that they produce qualitatively incorrect results, so that their applicability to a given problem has to be established first [67].

H_{ee} is the force field used to describe the environment. It is not normally different from other force fields. Particular attention has to be given to H_{pe}, the interaction between the primary system and environment. It is comparatively straightforward if the primary system and the environment are not connected by chemical bonds (e. g., the environment consists of separate solvent molecules). In this case, the primary and environment atoms interact through nonbonded force field terms, e. g., Lennard-Jones. In addition, the partial charges of the environment atoms may enter the one-electron Hamiltonian for the electrons of the primary system so that polarization of the primary system is automatically accounted for. Examples of such a partitioning are dimethoxyethane in aqueous solution [52] and a nucleophilic substitution reaction of chloroamine in dimethyl ether [68]. If the environment is linked to the primary system through chemical bonds, the embedding is more complicated. Typically, the dangling bonds of the primary system are saturated with hydrogen atoms and the interaction between the first few atoms on the primary side and the first few atoms on the environment side is handled exclusively by a force field which often needs special adaptation. An example of such a partitioning would be an enzyme reaction where the substrate and the active site residues constitute the primary system, and the rest of the enzyme plus water and counterions the environment. The active site residues are obviously connected to the rest of the protein. Using such a scheme, it has been possible to elucidate the mechanism of HIV protease [69].

2.4
Adaptive Interpolation of Potential Energy Surfaces

As stated in Sect. 2.3, hybrid schemes suffer from the dilemma that using semiempirical methods gives a wrong description of the primary system whereas ab initio methods are so expensive that the affordable number of time steps does not allow proper statistical-mechanical sampling. For selected systems, however, there is a possibility of using very accurate methods and still executing many time steps. The requirements on the system are that the internal conformational space of the primary system is finite and that the primary system can be described by only a few active degrees of freedom.

For such systems, it is possible to use adaptive interpolation schemes for the potential energy surface. The idea is as follows: When the system reaches a certain configuration for the first time its energy and forces are calculated quantum-chemically. In contrast to standard molecular dynamics, this information is not discarded after the next time step but is stored. Using these "old" conformations, a mesh is developed for the potential energy surface. If the system, at a later stage of the simulation, returns to this area of conformation space, its energy and forces are now interpolated from neighboring mesh points. Building up the interpolation mesh automatically during a simulation has the advantage that it will be densest in areas of low potential energy which are likely to be revisited by the system and where the system will spend most of the time (Boltzmann factor).

Both regular and irregular (finite element) meshes have been used [70,71]. Although the method appears to be quite straightforward many technical problems have had to be overcome to make it stable and efficient. It has recently been applied to the photoisomerization of stilbene in its first excited state in solution (for a review, see [72]). The stilbene (1,2-diphenylethene) molecule has the necessary properties. To a first approximation all its degrees of freedom can be treated as rigid except three dihedral angles (central C–C bond plus two dihedrals defining the orientation of the rigid phenyl rings). This gives rise to a three-dimensional potential energy surface that needs to be interpolated. Moreover, the conformational space is finite, since the dihedral angles can only take values between 0 and 360°. The use of the mesh allowed the study of conformational dynamics in the excited state starting from both the cis and the trans conformer using ab initio configuration interaction (CI) calculations [71]. So far, the treatment is at the level of singly-excited CI in a 6-31G* basis set. This is certainly not the ultimate level one would use. It is, however, a vast improvement over previous semiempirical treatments.

2.5 Ab Initio and Car-Parrinello Molecular Dynamics

The most rigorous way of incorporating quantum information into molecular simulation is, of course, by evaluating energies and forces quantum-chemically for the whole system during a simulation. Ab initio methods are usually limited to systems of a few atoms only and to short simulation times (picoseconds). Typical examples are the photodissociation of formaldehyde into hydrogen and carbon monoxide ($H_2CO \rightarrow H_2 + CO$) in vacuum, which has been studied at the HF/6-31G** level [73], and the rearrangement of small sodium clusters (Na_4) of various electronic states in vacuum, which was studied by the generalized-valence-bond method using effective core potentials [74]. More promising for macromolecular systems is the Car-Parrinello method. It is faster than traditional ab initio quantum chemistry due to several features: (i) Use of a dynamical scheme to adjust the electronic coefficients rather than diagonalization of the Hamiltonian. (ii) Plane waves rather than Gaussian basis sets. (iii) Effective core potentials (pseudopotentials) rather than all-electron calculations. (iv) Density functionals for Coulomb, exchange and correlation interactions. Using massively parallel computers, Car-Parrinello calculations can now be used to simulate systems of several dozen atoms for several dozen picoseconds. Successful applications include the dissociation of liquid water [75] and catalysis in zeolites [76]. More recently, the Car-Parrinello method and its derivatives have been employed directly in polymer science. Röthlisberger et al. [77] have described the mechanism of the anionic living polymerization of isoprene. Montanari and Jones have used density-functional techniques to describe crystalline polyethylene [78].

3 Efficient Generation of Dense Polymer Systems by Structural Simplification

3.1 Motivation

All the bulk properties of polymers are dependent on a number of different factors, prominent among them the chemical structures as well as the arrangement of the macromolecules in a dense packing [1–6]. Since it is very difficult to completely change the starting structure during the simulation of dense atomistically detailed polymer systems, one should start with a reasonably "good" structure. The problem consists of creating a packing of chains with conditions of spatial continuity where the "parent chains", belonging to the "original box", do not intersect themselves and also do not intersect their periodic images. The problem of generating dense polymer systems is formidable due to the high density and the connectivity of polymer systems. For crystal structures this can be systematically achieved [33–34]; for amorphous structures, however, there is no generally satisfactory method available.

Much effort has been expended constructing "reasonable" amorphous packings starting from atomistic models of single polymer chains. Common approaches can be roughly divided into three groups: the first comprises methods that grow chains into dense phases by adding new segments according to a Metropolis criterion [79–82], the second group starts from coarse "initial guesses" created on a lattice [27,28,30,83], and the third departs from structures with very low densities that are then condensed step by step during NpT-MD simulation to experimental densities of polymer systems [84]. An alternative technique starts by preparing the monomer liquid in the periodic box and then polymerizing the monomers to form a chain [85]. Common to these methods is that the "initial guess" structures are subsequently relaxed by potential energy minimization or simulated annealing where the temperature is gradually reduced to the temperature of interest.

Today, the problem still lies in the ad hoc nature of the initial-guess generation. The structures generated simply do not provide an ensemble in the statistical-mechanical sense. The quality of these structures is very much a function of the constructor and the care in testing. Nevertheless, today, ad hoc structures can be used to advantage in mechanistic investigations of macroscopically isotropic amorphous structures. In recent years many atomistic simulations using ad hoc structures have been carried out in the area of polymer-guest systems, most of which have focused on the calculation of the diffusion coefficient and transport phenomena of simple gases in amorphous polymers [86–95]. We will review here two recent developments in methods for generating amorphous packing.

3.2
Dense Configurations of Linear Chains of Constant Curvature

The overall static properties of flexible polymers can be deduced from the geometry of their backbone, if properly parameterized for their chemical structure [22,96,97]. Considering linear chains without a chemical structure tremendously simplifies the nature of the polymer configuration to be generated. The loss of information about the chemical details of the macromolecules, however, has to be recovered, in a later stage, by reconstructing the atomistically detailed chain. This procedure of "inverse mapping" has been successfully applied to lattice models [27,28,30,83] and can be easily extended to more complicated polymers. Recently an alternative approach has been proposed [98], in which the mesoscopic properties of macromolecules, such as the stiffness of the chains and scaling behavior, are maintained during a dramatic structural simplification.

3.2.1
From Atomistic Detailed Polymers to a Continuous Model

A novel approach [98], proposed for generating starting configurations of amorphous dense polymeric systems, departs from a continuous vector field and its stream lines. The stream lines of continuous vector fields never intersect. If the backbones of linear polymer chains can be associated with such stream lines, the property of the stream lines partly alleviates the problem of excluded volume, which – due to high density and connectivity – constitutes the major barrier to an efficient packing method of dense polymeric systems. This intrinsic "repulsive contact" can be compared to an athermal hard-core potential. Considering stream lines immensely simplifies the problem.

Since a finite segment of a stream line can start at every location of a vector field, the choice of the starting points for a set of stream lines alone determines their configuration. By simply choosing different starting points, it is possible to generate different configurations of linear chains – the stream lines – from a given vector field. As it turns out that the "excluded volume" condition is fulfilled for any number of stream lines, the choice of density for the stream lines to be generated can be made independently of the vector field.

The stream lines of a plain vector field, however, do not in general have a polymer-like behavior. The problem, then, consists of generating a vector field whose stream lines have the mesoscopic properties of real polymers.

In order to investigate a continuous vector field, one first needs to parameterize this continuum of vectors. Since one is interested in the stream lines, only unit vectors indicating the direction of propagation of the stream lines are relevant. Thus, it is sufficient to employ a set of spatially varying polar coordinates $\phi(\mathbf{x}) \in [0, \pi)$ and $\varphi(\mathbf{x}) \in [0, 2\pi)$ to describe the vector field. The vector field can be written as

$$\mathbf{v}(\mathbf{x}) = \begin{bmatrix} \sin\phi\cos\varphi \\ \sin\phi\sin\varphi \\ \cos\phi \end{bmatrix}, \quad \begin{array}{l} \phi(\mathbf{x}) \in \mathcal{R} \text{ modulo } \pi\mathbf{Z}, \\ \varphi(\mathbf{x}) \in \mathcal{R} \text{ modulo } 2\pi\mathbf{Z} \end{array}. \tag{3.1}$$

Note that this choice implies that the vectors always point in a well-defined direction, avoiding singularities of the stream lines. So far, the vector field is given in terms of fields $\phi(\mathbf{x})$ and $\varphi(\mathbf{x})$, which are not yet known. Thus, one needs to express these two scalar fields in subsumable discrete coordinates. Various complete sets of basis functions $\{\chi_j, j \in I\}$ exist, but for the current purposes only periodic functions in the simulation box Γ can be used (Γ denoting the periodic box of the atomistically detailed system). Fourier functions have the added advantage of conforming to well-defined length scales. Denoting the expansion coefficients of $\phi(\mathbf{x})$ and $\varphi(\mathbf{x})$ in this basis set by $\{a_j^{\varphi}, j \in I\}$ and $\{a_j^{\phi}, j \in I\}$, respectively, the scalar field $\phi(\mathbf{x})$, for instance, reads

$$\phi(\mathbf{x}) = \sum_{\mathbf{n}=(k,l,m)}^{|k|,|l|,|m|\leq N_\lambda} a_{\mathbf{n}}^{\phi} \cdot \exp\left(\frac{2\pi \mathrm{i}}{\mathrm{L}_\Gamma} \mathbf{n} \cdot \mathbf{x}\right) \tag{3.2}$$

where L_Γ denotes the linear size of the simulation box Γ, N_λ the linear cutoff in the mode decomposition of the scalar field, and i the imaginary unit. For a periodic simulation box of unit side length, the reciprocal space of the Fourier mode is a three-dimensional cubic grid, the lattice points of which have integer components ranging between $-N_\lambda$ and N_λ. Since the vectors of the vector field only have real components, the effective number of independent Fourier coefficients is restricted by the additional conditions that $\overline{a_{\mathbf{n}}^{\phi}} = a_{-\mathbf{n}}^{\phi}$ and $\overline{a_{\mathbf{n}}^{\varphi}} = a_{-\mathbf{n}}^{\varphi}$.

The stream lines of a vector field $\mathbf{v}(\mathbf{x})$ are those trajectories where the vector $\mathbf{v}(\mathbf{x})$ is tangential to the path. In analogy to trajectories of atoms subject to the influence of a Hamiltonian, the stream lines obey an "equation of motion" of first order given by

$$\frac{d}{ds}\mathbf{x}(s) = \mathbf{v}(\mathbf{x}(s)) \tag{3.3}$$

the solution of which is uniquely determined by the starting point $\mathbf{x}(s)|_{s=0} = \mathbf{x}_0$. The major difference to equations of motion of atoms, however, lies in the fact that the vector field does not need to be defined as the gradient of a scalar function. Note that, since the vector field has been chosen so that $\|\mathbf{v}(\mathbf{x})\|^2 = \mathbf{v}(\mathbf{x}(s)) \cdot \mathbf{v}(\mathbf{x}(s)) = 1$, one automatically obtains

$$\frac{d}{ds}\mathbf{v}(\mathbf{x}(s)) \cdot \mathbf{v}(\mathbf{x}(s))\Big|_\Gamma \equiv 0, \tag{3.4}$$

over the entire domain Γ on which the vector field is defined, which means that the changes of direction of the stream lines are always perpendicular to the stream lines themselves. This property will be used later.

The vector field entirely and uniquely determines the stream lines and their properties. As we focus our attention on the mesoscopic properties of stream lines, assuming that they can resemble a polymer-like amorphous packing of chain backbones, we have to consider in greater detail their intrinsic properties. As shown in the next section, Santos and Suter [98] elaborated a model for generating packing structures of Porod-Kratky chains.

3.2.2
Dense Configurations of Porod-Kratky Chains

The most austere representation of a polymer backbone considers continuous space curves with a persistence in their tangent direction. The Porod-Kratky model [99,100] for a chain molecule incorporates the concept of constant curvature c_0 everywhere on the chain skeleton; c_0 being dependent on the chemical structure of the polymer. It is frequently referred to as the wormlike chain, and detailed studies of this model have already appeared in the literature [101–103]. In his model, Santos accounts for the polymer-like behavior of stream lines by enforcing this property of constant curvature.

The stream line of a continuous vector field is a regular curve, the tangential vector of which is given by the vector field itself. A regular curve is a piecewise-differentiable function $\mathbf{r} : [a, b] \subseteq \mathbf{R} \rightarrow \mathcal{R}^3$ and its "velocity" $\dot{\mathbf{r}}(s) = \partial/\partial s\mathbf{r}(s)$ is nonzero everywhere, i. e., $\|\dot{\mathbf{r}}(s)\| \neq 0$ for $s \in [a, b]$. The "motion" of a triplet of orthonormal vectors comoving along the curve is entirely described by the Frenet formulae [104]. In the plane of the tangential circle of radius ρ, the curvature $c(s)$ of the curve $\mathbf{r}(s)$ is given by

$$c(s) \equiv \frac{1}{\rho} = \frac{\|\dot{\mathbf{r}}(s) \times \ddot{\mathbf{r}}(s)\|}{\|\dot{\mathbf{r}}(s)\|^3}. \tag{3.5}$$

Note that the curvature is independent of the kind of parameterization s of the curve, which indicates that, in the case of a stream line, the curvature can be made to depend on the location of this point in the vector field and not on the parametrization along the stream line itself. By virtue of Eqs. (3.3), (3.4), and (3.5) and some manipulations, the curvature of a stream line at a location $\mathbf{x}$ can be expressed as:

$$c(\mathbf{x}) = \left\| \frac{d}{ds}\mathbf{v}(\mathbf{x}(s)) \right\|. \tag{3.6}$$

Equation (3.6) defines a scalar field, the value of which at any point in the simulation box is the curvature of the stream line passing through this point and can also be reformulated to avoid the use of the parameter s:

$$\frac{d}{ds}\mathbf{v}(\mathbf{x}(s)) = \lim_{\epsilon \to 0} \frac{\mathbf{v}(\mathbf{x} + \epsilon\mathbf{v}(\mathbf{x})) - \mathbf{v}(\mathbf{x})}{\epsilon} = \mathbf{v}(\mathbf{x}) \cdot \nabla\mathbf{v}(\mathbf{x}). \tag{3.7}$$

Here it is most convenient to introduce an orthonormal set of unit vectors:

$$\mathbf{e}_r(\mathbf{x}) = \frac{\mathbf{v}(\mathbf{x})}{\|\mathbf{v}(\mathbf{x})\|}, \quad \mathbf{e}_\phi(\mathbf{x}) = \frac{\partial}{\partial\phi}\mathbf{e}_r, \quad \mathbf{e}_\varphi(\mathbf{x}) = \frac{1}{\sin\Theta}\frac{\partial}{\partial\varphi}\mathbf{e}_r,$$

and this vector basis yields the essential term in Eq. (3.7) as

$$\frac{d\mathbf{v}}{ds} = \mathbf{v}_k\partial_k\mathbf{v} = \mathbf{v}_k\partial_k\phi \cdot \mathbf{e}_\phi + \sin\phi\mathbf{v}_k\partial_k\varphi \cdot \mathbf{e}_\varphi, \quad k \in \{x, y, z\}, \tag{3.8}$$

which simply states that the infinitesimal changes in this tangential vector along a stream line are perpendicular to the vector $\mathbf{e}_r(\mathbf{x})$, the direction of the stream line. Hence, the spatial dependency of the scalar curvature field is

$$\begin{aligned} c(\mathbf{x})^2 &= \|\partial_k\phi\mathbf{v}_k \cdot \mathbf{e}_\phi + \sin\phi\partial_k\varphi\mathbf{v}_k \cdot \mathbf{e}_\varphi\|^2 \\ &= (\partial_k\phi\mathbf{v}_k)^2 + (\sin\phi\partial_k\varphi\mathbf{v}_k)^2. \end{aligned} \tag{3.9}$$

This curvature field is always positive and has no singularity for continuous differentiable fields $\phi(\mathbf{x})$ and $\varphi(\mathbf{x})$. Given these two fields, the stream lines can be constructed by integrating Eq. (3.3), starting from any set of freely chosen points in the simulation box.

In the case of the Porod-Kratky model, the polymer backbones have a constant curvature c_0. Accounting for the polymer stiffness in generating the dense configuration of stream lines, the vector field used must have a homogeneous curvature field with a unique value c_0 in the entire simulation box Γ. In order to quantify the success in creating such a vector field, the deviation of the curvature from the ideal Porod-Kratky case, a volume integral has been used by Santos as a penalty function:

$$\overline{E}\left(\{a_j^\varphi, a_j^\phi\}\right) \propto \int_\Gamma dx^3(c(x) - c_0)^2. \tag{3.10}$$

The error function $\overline{E}$ explicitly depends on the set of coefficients $\{a_j^\varphi, a_j^\phi, j \in I\}$ that parameterize the vector field and has the major advantage of having a unique minimum of zero in the ideal Porod-Kratky case.

So far, Santos has been able to express the relation between a set of coefficients $\{a_i^\varphi, a_j^\phi, j \in I\}$ describing a vector field and the overall curvature of the stream lines of this vector field. Based on the curvature field, they constructed the measure $\overline{E}$ of the curvature distribution in the simulation box. Provided that the homogeneous curvature field of curvature c_0 is the one that minimizes $\overline{E}$, the problem of packing has been recast as a minimization problem. However, the lack of information about the gradient of the error function to be minimized does not facilitate the search. Fortunately, appropriate computer simulation schemes for similar minimization problems have been proposed in the literature [105–109].

3.2.3
Computational Method

Some approximations are required. First, the evaluation of the integral in Eq. (3.10) within the simulation box with periodic continuation conditions is approximated by a sum of local contributions. This sum is calculated over N_p test points $\{\mathbf{x}_k, 1 \leq k \leq N_p\}$ which are homogeneously distributed on a three-dimensional grid covering the entire simulation box and is given by:

$$E\left(\{a_j^{\varphi}, a_j^{\phi}\}\right) = \sum_{k}^{N_p} (c(\mathbf{x}_k) - c_o)^2 . \tag{3.11}$$

Second, having to consider only a finite number N_λ^3 of coefficients $\{a_j^{\varphi}, a_j^{\phi}, j \in I\}$ correspondingly restricts the number of basis functions, i. e., of Fourier modes in Eq. (3.2); these modes encompass the essential features of the vector field.

As a consequence of the aforementioned discretization, the number of test points N_p, and the linear cutoff N_λ in the mode decomposition turn out to be closely related. Denoted by $\Delta\mathbf{x}$, the typical linear distance between two adjacent test points, $\Delta\mathbf{x}$ must be small enough to ensure that the integral function $\overline{E}$ is well approximated by the sum E, i. e., $\Delta\mathbf{x}$ must fulfil the condition $c(\mathbf{x} + \Delta\mathbf{x}) \approx c(\mathbf{x})$. This is to say that the length scale $\Delta\mathbf{x}$ directly determines the minimal length scale contribution of the Fourier modes, which is $\mathrm{L}_\Gamma / N_\lambda$. Consequently, after having fixed N_λ, the spatial variation of $c(\mathbf{x})$ as a function of the number of test points N_p has to be carefully monitored to determine the maximum distance $\Delta\mathbf{x}$ ensuring that $c(\mathbf{x} + \Delta\mathbf{x}) \approx c(\mathbf{x})$ in the entire box. Typically, for $N_\lambda = 8$, we have found that a minimum of 23^3 test points in a box of unit length is necessary.

The minimization problem consists of zeroing the discretized function in Eq. (3.11) in the space of coefficients $\{a_j^{\varphi}, a_j^{\phi}, j \in I\}$. Santos used a simulated annealing method [106–114], which employs a crude analogy to an adiabatic cooling process by means of a properly tuned Monte Carlo scheme [105]. The stationary state of a physical system, closed but in contact with a thermal bath of temperature T, obeys the Gibbs distribution. In the limit $\mathrm{T} \rightarrow 0$, only the configurations of minimal energy can be occupied; this is the ground state of the system. Such a final distribution can only be achieved by cooling carefully keeping it in quasi-equilibrium at each step. An advantage of such a method is a gain in a better understanding of the topological nature of the energy hypersurface the system can move on [109].

An appropriate choice of the simulated annealing algorithm for the Monte Carlo method is a prerequisite to ensuring that the stochastic sequence converges asymptotically towards a stationary distribution. The freezing scheme has to be very slow and some theoretical (but unrealistic) lower boundary for the decrement rule for the temperature has been derived, guaranteeing convergence in a very large but finite number of steps [110]. In practice, one must define a cooling schedule by specifying the initial value of the temperature, the final value of the temperature, and a decrement rule for the temperature. Santos and Suter opted for the following standard annealing scheme [109]

$$T = T_0 + (T - T_0)\left(1 - \frac{t}{t_{max}}\right)^{\alpha},$$

where the exponent $\alpha \in [1, 4]$ and t measures the numbers of simulation steps.

In addition to the influence of the temperature, the acceptance rate of the Monte Carlo simulation is controlled by the maximal amplitude of changes in the sets of coefficients $\{a_j^{\varphi}\}$ and $\{a_j^{\phi}\}$ for the Monte Carlo moves. An equilibration phase (10 to 20% of the time spent at each temperature niveau) has to be carried out and during this equilibration, the maximal amplitude of change of the coefficients $\{a_j^{\varphi}\}$ and $\{a_j^{\phi}\}$ has to be tuned to match the target values 30 to 50%. After this, it is crucial to fix these amplitudes for the rest of the time spent at each temperature niveau since only then is the microreversibility of the Monte Carlo process ensured, a prerequisite for the convergence to the canonical distribution of the configurations of the system during the runs. This cooling schedule is shown in Fig. 3.1.

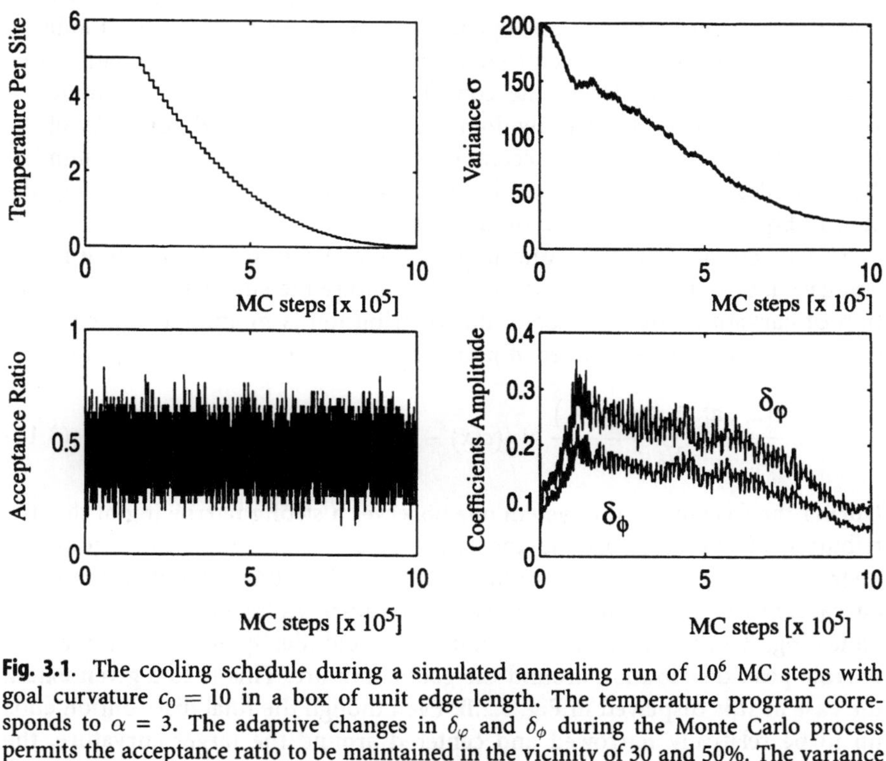

Fig. 3.1. The cooling schedule during a simulated annealing run of 10^6 MC steps with goal curvature $c_0 = 10$ in a box of unit edge length. The temperature program corresponds to $\alpha = 3$. The adaptive changes in δ_{φ} and δ_{ϕ} during the Monte Carlo process permits the acceptance ratio to be maintained in the vicinity of 30 and 50%. The variance s of the distribution of curvature is monitored along the annealing process

3.2.4
Results and Discussion

Santos [98] carried out the simulated annealing simulation on DEC 8400 machines. For the model system comprising 17^3 ($N_\lambda = 8$) Fourier modes and 23^3 test points in a simulation box, on average 16 MC steps per second could be performed. The length of the runs extends from $2 \cdot 10^6$ to $5 \cdot 10^6$ MC steps.

The model system is a periodic box of arbitrary unit side length. A linear cutoff $N_\lambda = 8$ in the frequency spectrum of the Fourier decomposition corresponds to a minimal characteristic length $\lambda = 0.125$ for the scalar fields $\phi(\mathbf{x})$ and $\varphi(\mathbf{x})$. The investigated systems have goal curvatures c_0 chosen from the set $\{0.1, 0.2, 0.5, 1, 5, 10\}$.

Every minimization departs from an initial estimation for the vector field. The minimizations were carried out with a starting configuration obtained by randomizing the coefficients $\{a_j^\varphi\}$ and $\{a_j^\phi\}$; the resulting vector field has no preferential orientation and the distribution of curvature in the simulation box exhibits a long tail mainly due to abrupt changes in the direction of the stream lines (see Fig. 3.2A).

The acceptance criterion for the Monte Carlo scheme is simple Glauber dynamics. For computational reasons, a single Monte Carlo move comprises a change in only one of the Fourier modes. A very efficient calculation scheme underlies this choice of a "local- update" scheme: the sum of Fourier modes in Eq. (3.2) do not need to be entirely recalculated at each MC step, but can be obtained by solely updating the contribution of the one Fourier mode involved in the single MC step. By doing so, the computational efficiency of the simulated annealing scheme becomes quite insensitive to the linear number N_λ of modes. Most of the elapsed simulation time is spent in evaluating the sum in Eq. (3.11) over the test point grid.

The effects of simulated annealing on the distribution of curvature in the simulation box are shown in Fig. 3.2.B. To capture the changes in the distribution of curvature, the variance of this distribution is monitored, which turns out to be – for a large number of points N_p – very well approximated by

$$\sigma^2 \equiv \frac{E\left(a_j^\varphi, a_j^\phi, 1 \le j \le N_\lambda\right)}{N_p} \approx \langle (c(x) - c_0)^2 \rangle . \tag{3.12}$$

During the freezing of the system one observes a strong narrowing of the distribution of curvature around the target value c_0. Starting from a long tail distribution, the initial equilibration phase already suffices to reduce the mean curvature to a small multiple of the goal curvature. At the end of the simulated annealing, the final curvature distribution should ideally be reduced to a very narrow peak distribution around the goal curvature. This is to a given extent the case as it is displayed in Fig. 3.2.B. Even though the final distribution ends up to be relatively narrowed and centered around the target curvature, the shape of the distribution is strongly dependent on the goal curvature (see

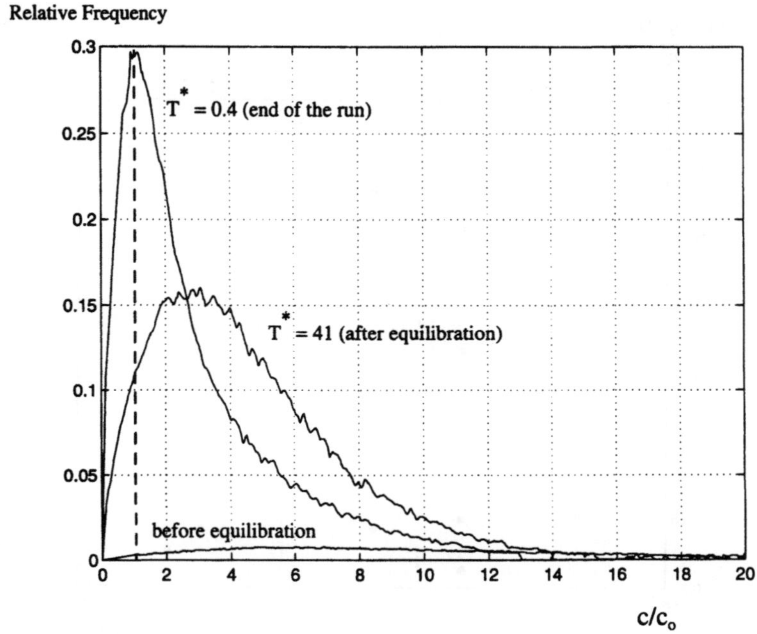

a

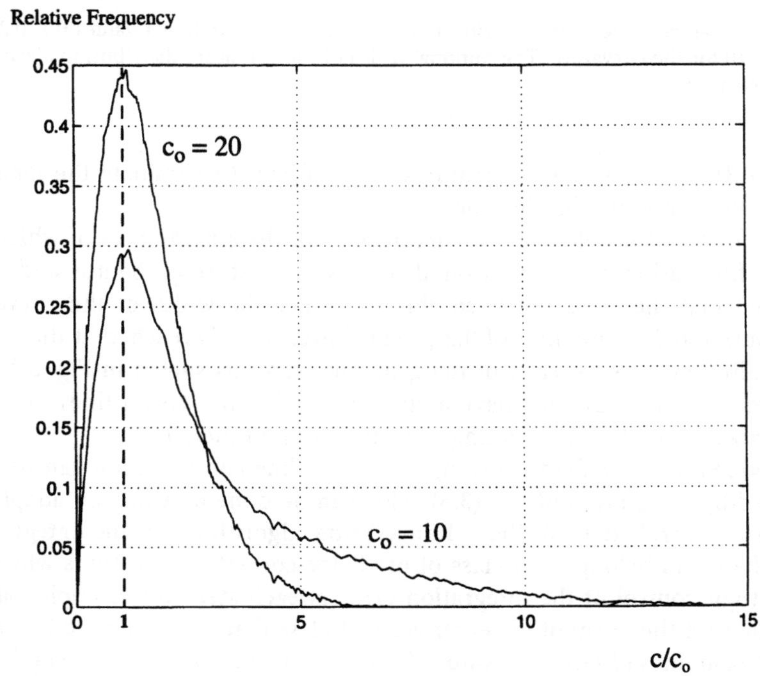

b

Fig. 3.2. **a** Distribution of curvature in the simulation box during the three stages (random starting configuration, after equilibration and at the end of simulated annealing) of a simulated annealing run with $c_o = 10$. T^* denotes the temperature. **b** Final distribution of curvature for two different goal curvatures c_o.

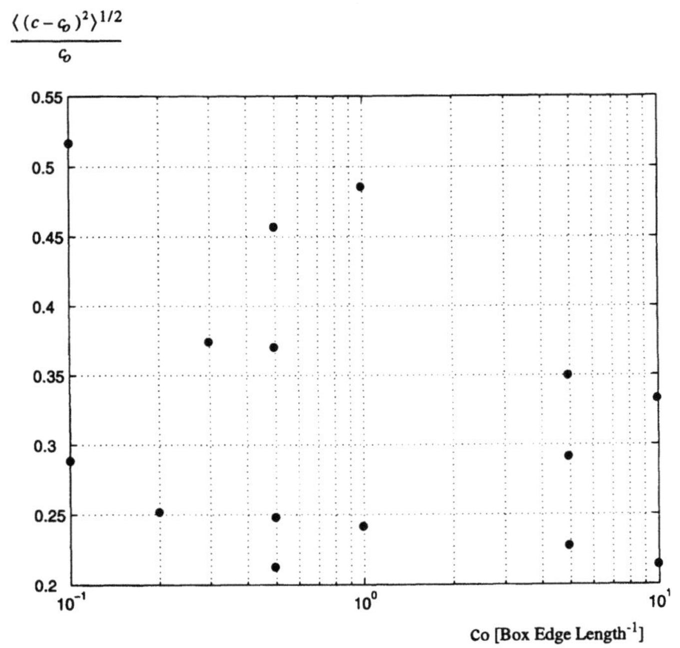

Fig. 3.3. Final variance of the distribution of curvature for the target values 0.1, 0.2, 0.5, 1, 5 and 10 for the curvature. The minimal width of the curvature distribution always lies just above $0.2c_o$

Fig. 3.2.B). For small goal curvatures it is difficult to eliminate the long-tail part of the curvature distribution.

The final vector fields have a curvature field, the stream lines of which exhibit a fluctuating curvature around the goal curvature c_0. Santos and Suter reported that they had not been able to reduce the width of the curvature distribution below the limit of $0.2c_0$. This lower limit is reached at the end of the cooling process for each of the goal curvatures as shown in Fig. 3.3. The final vector fields obtained have a curvature field, the stream lines of which are characterized by a fluctuating curvature around the target c_0.

Provided a vector field is given, the stream line configuration can be constructed by integration of Eq. (3.3). This can be done by using an adaptative step-size controlled Cash-Karp Runge-Kutta algorithm for the integration. This algorithm belongs to a class of predictor-corrector algorithms which allow a tight control of the integration error at each step (for example, Santos and Suter set the norm of the error vector to less than 10^{-9}). Figure 3.4 shows four stream lines of target curvature ($c_0 = 10$) in a box of unit side length. The length of the curve is twelve times the box side length so that the projection onto the xy, yz, and xz planes of the stream lines extend out of the simulation box. The curvature along these four stream lines has been calculated by finite difference and is depicted in Fig. 3.5.

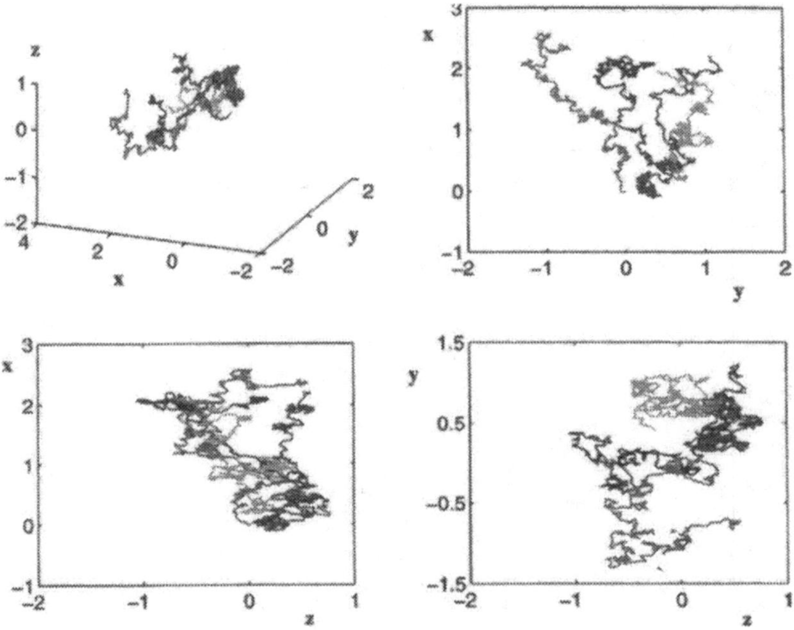

Fig. 3.4. Four stream lines and their projection onto the xy, yz and xz planes, respectively, constructed by integration for a vector field of curvature $c_o = 10$ in a box of unit edge length

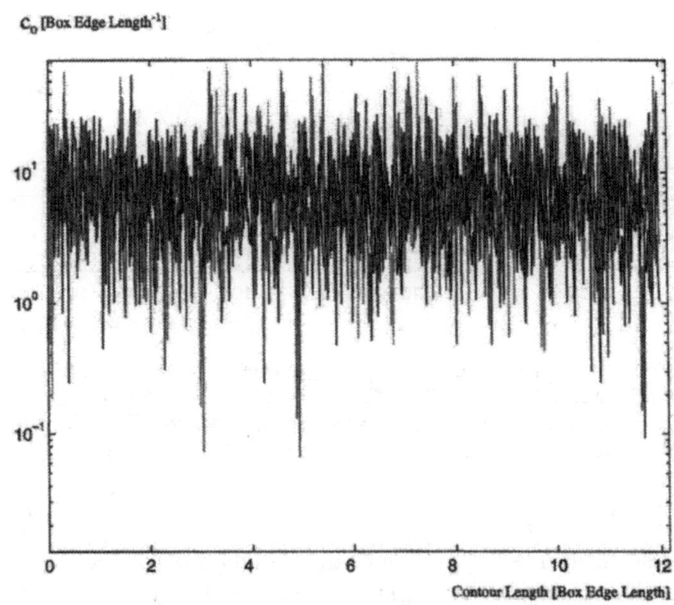

Fig. 3.5. Curvature along the four stream lines depicted in Fig. 3.4 with target curvature $c_o = 10$ for the underlying vector field

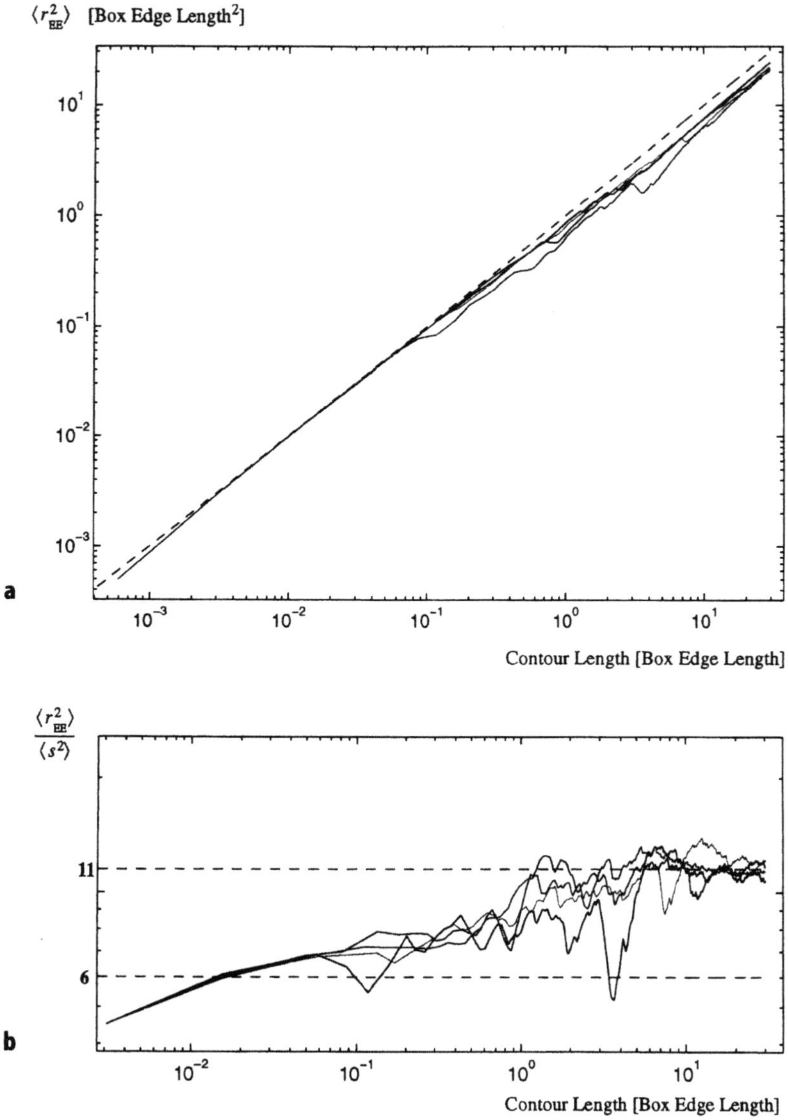

Fig. 3.6. **a** End-to-end distance of the four stream lines depicted in Fig. 3.4 as a function of the contour length of the stream lines. **b** Ratio of the end-to-end distance to the square of the radius of gyration as a function of the contour length of the stream lines. The Porod-Kratky value of 6 is achieved only for a contour length of the same order as the box edge length

Santos also scrutinized the polymer-like behavior of the constructed stream lines. Two characteristics were analyzed: the second moment of the end-to-end distance distribution, $\langle r^2 \rangle_0$, which characterizes the spatial configurations of chain molecules, and the radius of gyration, R_G, that indicates how the backbone elements are distributed around the center of mass of the chains. In the case of Porod-Kratky chains, $\langle r^2 \rangle_0$ is expected to be proportional to the length of the chain backbone for large chain length, which is clearly the case (Fig. 3.6.A). The theoretically expected ratio $\langle r^2 \rangle_0 / R_G^2$ is 6 for Porod-Kratky chains. The stream lines correctly reproduce this ratio only to contour length of the same order as the box edge length (Fig. 3.6.B). For a much higher contour length, the radius of gyration is half as large as expected.

3.2.5
Conclusions

The results from simulated annealing combined with an integration scheme demonstrates the feasibility of a vector field method. Once the proper vector field has been generated, several configurations of any desired density of linear chains can be generated with ease. While the computational efficiency of the minimization scheme is not sensitive to the number of modes, attention must be paid to the choice of the number of test points in the simulation box; a large number of points requires lengthy calculations.

We have not mentioned here the crucial inverse mapping of the realistic polymer structure onto the stream line. For polymers without side groups, such as polyethylene or bisphenol-A-polycarbonate, the following strategy has successfully been used [98]: an energy minimization of the internal energy contributions was carried out simultaneously with a minimization of the distances of all atoms to the stream line (to this end, the sum of the squared dis-

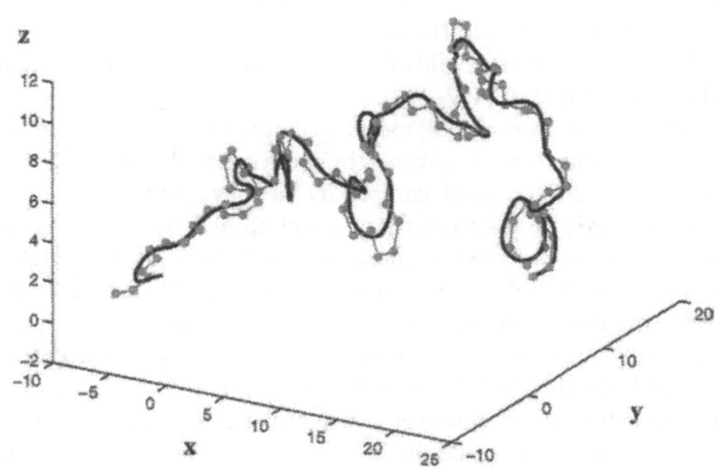

Fig. 3.7. A united-atom PE chain constituted of 100 "united atoms" mapped around a stream line. The bond angles as well as the bond lengths have been kept fixed. The torsion angles have been subject to a minimization considering a three-fold potential accounting for the simplified chemical structure of the chain. An additional "proximity function" has been used to force the chain to follow the trajectory of the stream line [114]

tances of all atome from the stream line was added – with a scaling factor – to the total energy of the system). This stage was then followed by a global-energy minimization. An united-atom-polyethylene chain successfully mapped around a stream line is shown in Fig. 3.7.

3.3
A Novel Algorithm Based on a Parallel-Rotation Move for Generating Polymer Structures Efficiently

The problem of generating "good" starting configurations of atomistically detailed polymers can be tackled in many different ways. A new approach has recently been presented [115] that combines an explicit steepest-descent search algorithm in the space of generalized coordinates with an adaptive scheme for gradually incorporating the effect of the nonbonded interactions between the chains and their hard surrounding. The embedding problem incorporates most information about the local energy, due to the chemical structure of the polymer chain, by the way of torsion angle correlations provided either by the RIS model or by experimental measurements in the bulk polymer.

3.3.1
The Embedding Problem

The simplest formulation of the packing problem is to give some collection of distance constraints and to calculate these coordinates in ordinary three-dimensional Euclidean space for the atoms of a molecule. This embedding problem – the "Fundamental Problem of Distance Geometry" – has been proven to be NP-hard [116]. However, this does not mean that practical algorithms for its solution do not exist [117–119].

Fortunately, in the case of polymeric systems, the number of degrees of freedom may be considerably reduced by working with generalized coordinates [120–122]. The structural complexity of long polymer chains, which arises from geometry and connectivity, is simplified by assuming that the bond lengths and angles between consecutive backbone atoms remain fixed. Then the molecular geometry is uniquely determined by the successive torsion angles along the backbone and the position and orientation of the chain's start. This rigidly constrained model provides the relevant structural properties for atomistically detailed chains to be packed.

Nonbonded interactions, however, must also be considered. The relevant effect of the long-range interactions is primarily a repulsive contact interaction between the atoms of the macromolecules. Focusing their endeavour solely on this hard-core effect in generating a dense packing, Müller, Santos, Nievergelt, and Suter [115] assumed the atoms to be merely hard spheres, the hard-sphere radius being closely related to the van der Waals radius of the atoms [123]. The hard-sphere type of constraint has the major advantage of being a local constraint that is computationally very much less demanding than long-range nonbonded interactions.

Accounting properly for the chemically detailed structure of the macromolecules constitutes another prerequisite of the embedding problem. The rotational isomeric state (RIS) method [35–36,124] was created to make it possible to describe the conformational behavior of macromolecules with proper attention to the details of the chemical structure. Its application to macromolecules is easy, if one is willing to neglect the effects of excluded volume. The essential simplifying assumption of the RIS approach consists of the replacement of continuous geometrical degrees of freedom by discrete states, which are chosen so that the conformational partition function can be factorized into local contributions of these discrete states. In addition, factorization of the partition function provides a formidable vehicle for the efficient estimation of thermally averaged properties such as torsion angle distributions or correlation distributions of adjacent torsion angles along the chain backbone. A local-energy model for the conformers suffices for extracting the relevant information providing the average distribution by means of the RIS method [35–36].

Müller et al. [115] specify a number of discrete states for the torsion angles according to the rotational isomeric states of the considered macromolecule. The RIS model also gives very valuable information about the correlation of two adjacent torsion angles along the backbone chain. In their model these correlations constitute an additional collection of constraints, completing the distance constraints of the embedding problem. Experimental data, such as the recent NMR-measurements of angle correlations in polycarbonate [125,-126], could equally serve as geometrical constraints.

To solve the embedding problem basically requires the search in torsion angle space for a configuration for which none of the given geometrical constraints are violated (overlap of atoms) and for which the given torsion-angle correlations are satisfied. Solving this problem is tantamount to finding a procedure that simultaneously minimizes the intramolecular and intermolecular "thermal" energy without creating unphysical polymer structures. The complexity arising because of the many geometrical constraints together with the high dimensionality of the search space indicates that this kind of embedding problem is NP-hard [115]. Thus, the embedding algorithm proposed must be of heuristic nature and there is no guarantee that it will be successful in finding the proper configuration. Computer experiments, however, have shown that this algorithm converges in most of the cases towards a satisfactory solution.

3.3.2
The Search Algorithm

The major difficulty that the search algorithm has to face for long chains is that it is hardly possible to make any changes in the torsion angles without creating significant overlaps of atom pairs. Successful attempts to move several torsion angles without significant overlap are then made very improbable. Short polymer segments are clearly not subject to such strong limitations; this constitutes the clue to the solution of the embedding problem.

It appears that an efficient search algorithm beginning with short chain segments benefits from an enhanced mobility in the torsion angle space, but this does not help because the entire chain should be considered. Indeed, the connectivity of polymer chains has always been regarded as one of the bottlenecks rendering polymer simulation a difficult task. In order to circumvent the connectivity problem without cutting a long chain into segments that eventually will be put together again, we introduce the concept of a "horizon".

Consider a dense system of several long chains in a simulation box and define the "horizon" to be an arbitrary number of torsion angles. If a particular torsion angle of one of these chains has to be changed, only those atoms of this particular chain that are away from the changed angle by fewer torsion angles than the horizon can interact with each other, even though all the atoms are involved in the move. The "interchain" interactions accounting for the hard-sphere surrounding are only considered between pairs of atoms, which belong to the central segment of the chains that comprise as many torsion angles as the "horizon" value. For every "move attempt", the chain will be locally shortened by ignoring the atoms beyond the horizon (of course, all the atoms of the chain will be moved). Changing another torsion angle of the same chain results in a different fragment of the chain being analyzed. This is illustrated in Fig. 3.8. The horizon permits a remedy to the difficulty of moving chains in a dense phase.

The outer loop in the proposed algorithm will, thus, be a loop over the value of the horizon. Starting from a low value, the horizon is increased to

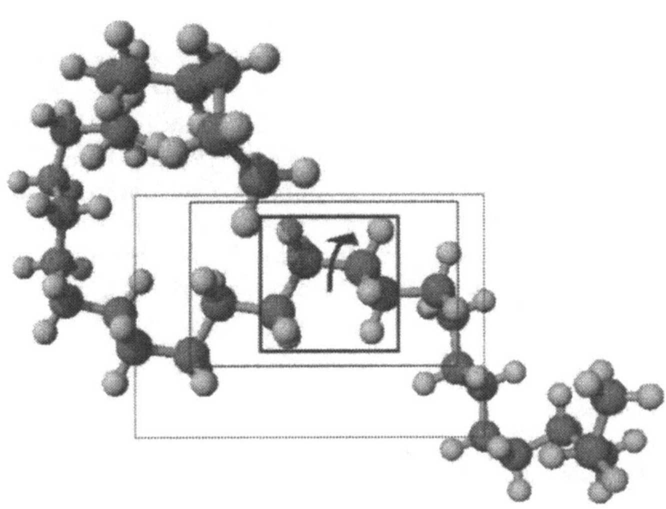

Fig. 3.8. Illustration of the concept of a "horizon"

reach its maximum value at the end of the simulation (the number of torsion angles of the chain backbone). In the interior of this main loop, an optimization loop serves to decrease the maximal overlap encountered in configurations during the run. Local collision is merely measured by the overlap of the hard spheres representing the atoms and is quantified for pair collisions of two hard spheres of radius r_1^0 and r_2^0, respectively, as

$$\mathrm{coll}_2(k,l) = \max\left\{0, \frac{(r_1^0 + r_2^0) - \|\mathbf{r}_k - \mathbf{r}_l\|}{r_1^0 + r_2^0}\right\} \tag{3.13}$$

where $\mathbf{r}_k$ and $\mathbf{r}_l$ are the position vectors of atoms k and l, respectively; note that $0 \leq \mathrm{coll}_2(k,l) \leq 1$. The "global collision" measure is defined as

$$\mathrm{coll}(h) = \max\{\mathrm{coll}_2(k,l) : \forall (k,l) \text{ within the scope of the horizon } h\}, \tag{3.14}$$

being the maximal encountered collision for pairs of atoms comprised within the scope of the horizon h. At each stage, with a fixed collision limit, the entire collection of torsion angles will be reviewed in a random manner. For each torsion angle, a move attempt will be made by exploring systematically the RIS states of this angle and by calculating the overlap induced by this move. Then the RIS state that simultaneously minimizes the collision coll(h) and satisfies the condition of correlation with its neighboring torsion angle is picked and the chain configuration is accordingly updated. This choice corresponds to a steepest-descent scheme. After all the torsion angles have been explored, the collision limit can be lowered further. However, if the maximal collision is larger than the collision limit, the collision limit itself will increase in order to allow configurations with large overlap to be temporarily accepted. This stage ends when the collision limit reaches a fixed lower boundary.

The reader will certainly have noticed that no mention has been made of the kind of torsion angle move used in the kernel of the iterative search process. In addition to the simplest move of one individual angle, the search scheme requires a robust and efficient move capable of involving all torsion angles along a chain segment (even the middle ones). Numerous off-lattice move schemes are available but, unfortunately, none of them has proved efficient enough for the desired search algorithm. A novel parallel rotation [127] move is a remedy.

3.3.3
A Novel Parallel-Rotation Move (ParRot)

Let us leave, for a moment, the domain of discrete torsion angles and consider torsion-angle moves in a torsion continuum. Much effort has been directed to the development of efficient phase-space sampling methods for MC [17,23,37,128–138]. The simplest off-lattice move – being a pivotal move [128,131] of a single torsion angle at a time – hardly samples configuration space, because of the bottlenecks produced by high density and chain entanglements. Reptation techniques have been reported [23,129,130,133,137] to be very ineffective for the treatment of long chains, since they rely on the existence of numerous chain ends. The configuration-bias-Monte-Carlo method (CBMC) [134–136,138] has attracted much attention, possibly, because of the very intuitive use of available "unoccupied" volume. As is the case with reptation, the CBMC method is found to be inefficient for the treatment of long chains in a dense system. The recently introduced concerted-rotation method [139,140] and its extensions [141] consists of moving seven consecutive middle torsion angles in a chain, keeping both residues of the chain around this moving unit fixed in space. Building such a "bridge" between two fixed sites has been demonstrated to be a well-posed geometrical problem [139]. Mainly due to a highly probable "geometric failure," this very involved sampling technique, however, does not constitute an improvement over existing methods [141]. To date, there has not been any conventional sampling technique robust and efficient enough to serve the purpose at hand.

A novel powerful off-lattice sampling technique, which has been demonstrated to be very suitable for treating long chains in the dense phase, has recently been proposed [127] by Santos et al. This parallel-rotation (ParRot) algorithm permits changing torsion angles that are deep inside long chains in a dense system. The ParRot move is extremely simple and computationally very efficient. It directly affects a moving unit of four consecutive torsion angles on the chain backbone, as shown in Fig. 3.9A, provided that the orientation of the parts of the chain outside the moving unit remain fixed. Moving four end sites of a hexamer involves a change in three torsion angles. The first torsion angle ϕ_0 is the driving angle for the move of the three other dependent angles, as demonstrated in Fig. 3.9.B. The orientation constraints are given by the vectors $\mathbf{u}_\perp$ and $\mathbf{u}$ that are coerced to remain constant for any value of ϕ_0. Both these vectors uniquely define the orientation of the remainder of the chain end to be moved. The following system of equations for the torsion angles $\{\phi_1, \phi_2, \phi_3\}$ are obtained:

$$\begin{cases} \mathbf{T}(\phi_0)\mathbf{T}(\phi_1)\mathbf{T}(\phi_2)\mathbf{e}_x = \mathbf{u} = \mathbf{T}(\hat{\phi}_0)\mathbf{T}(\hat{\phi}_1)\mathbf{T}(\hat{\phi}_2)\mathbf{e}_x \\ \mathbf{T}(\phi_0)\mathbf{T}(\phi_1)\mathbf{T}(\phi_2)\mathbf{T}(\phi_3)\mathbf{e}_y = \mathbf{u}_\perp = \mathbf{T}(\hat{\phi}_0)\mathbf{T}(\hat{\phi}_1)\mathbf{T}(\hat{\phi}_2)\mathbf{T}(\hat{\phi}_3)\mathbf{e}_y \end{cases} \tag{3.15}$$

where $\{\hat{\phi}_0, \hat{\phi}_1, \hat{\phi}_2, \hat{\phi}_3\}$ denotes the initial torsion angles. The rotation $\mathbf{T}(\phi) = \mathbf{R}^x_\theta \mathbf{R}^z_{\pi+\phi}$ (as defined by Mattice and Suter [36]) is composed of two rotations defining the local coordinate frames of a bond. Solving the first equation in Eq. (3.15) provides ϕ_2 expressed as:

$$\cos\phi_2 = \frac{\mathbf{v}_x(\phi_0, \hat{\phi}_0, \hat{\phi}_1, \hat{\phi}_2) - \cos\theta_1\cos\theta_2}{\sin\theta_1\sin\theta_2}, \tag{3.16}$$

for the vector $\mathbf{v}_x(\phi_0, \hat{\phi}_0, \hat{\phi}_1, \hat{\phi}_2) = \mathbf{T}(\phi_0)^{-1}\mathbf{T}(\hat{\phi}_0)\mathbf{T}(\hat{\phi}_1)\mathbf{T}(\hat{\phi}_2)\mathbf{e}_x$; θ_1 and θ_2 denote the bond angles between the bond of ϕ_0 and ϕ_1, and the bond of ϕ_1 and ϕ_2,

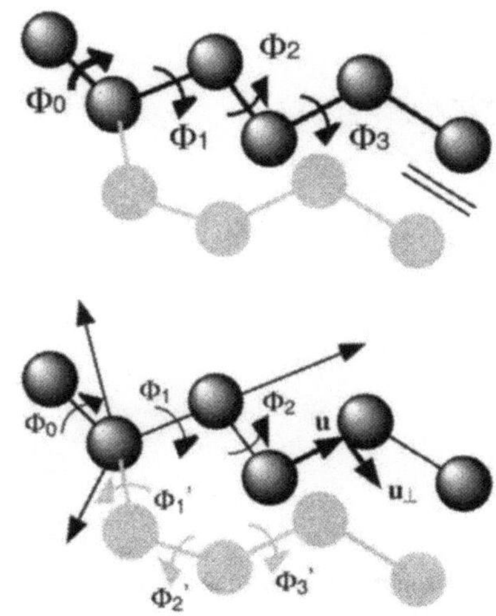

Fig. 3.9. **a** Illustration of the ParRot move and **b** The four torsion angles involved in a ParRot move that coerce the vectors $\mathbf{u}_\perp$ and $\mathbf{u}$ to remain constant for any value of ϕ_0

respectively. Equation (3.16) has either zero, one, or two solutions. Equation (3.15) permits the determination of the first torsion angle as:

$$\cos\phi_1 = \frac{a \cdot \mathbf{v}_y - b \cdot \mathbf{v}_z}{\sqrt{\mathbf{v}_x^2 - \mathbf{v}_y^2}} \tag{3.17}$$

where the coefficients are $a = \cos\theta_2 \sin\theta_1 - \sin\theta_2 \cos\theta_1 \cos\phi_2$ and $b = \sin\theta_2 \sin\phi_2$. The third torsion angle, ϕ_3, is responsible for the rotation of vector $\mathbf{u}_\perp$ around vector $\mathbf{u}$ that finally restores the original orientation along the direction set by the two angles ϕ_1 and ϕ_2 already found (see Fig. 3.10). Despite its relative simplicity, the ParRot move generates nontrivial three-dimensional trajectories for the moving chain residue, as depicted in Fig. 3.11.

Given the four initial torsion angles, the calculated amplitude of the displacements of the moving atoms to their original position clearly reveals that, despite large changes in the driving angle, at least one of the two possible ParRot-trajectories generally remains at a distance less than a bond length from its initial position (see Fig. 3.12.A). Thus, even in a dense system, the middle torsion angles of a backbone chain can undergo relatively large changes without "feeling" their hard environment. This becomes even more obvious in a NpT-MC simulation in a dense phase. Santos et al. have simulated a system of 10 polybead chains [142] of length 71 united atoms at a bulk density of 0.75 g/cm^3. As illustrated in Fig. 3.13, the ParRot-move can be performed at any depth in the chain. This is clearly not the case for the CBMC move. The ParRot-move is very appropriate for relaxing torsion angles even in dense systems of long chains.

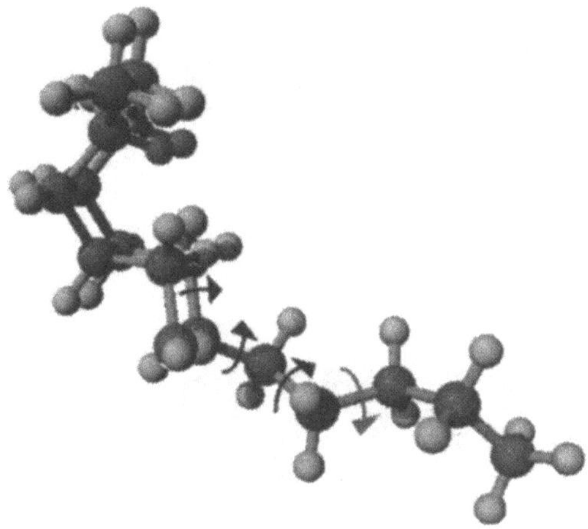

Fig. 3.10. Illustration of the ParRot move on a realistic polymer chain

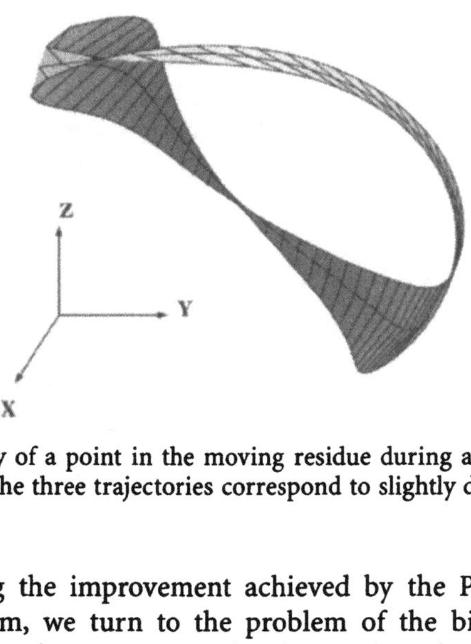

Fig. 3.11. 3D-trajectory of a point in the moving residue during an entire revolution of the driving angle ϕ_0. The three trajectories correspond to slightly different values for the fixed bond angles

Before examining the improvement achieved by the ParRot-move to the embedding algorithm, we turn to the problem of the bias induced by the move. ParRot involves four torsion angles that are interdependent. As a result, when departing from a uniform distribution of the driving torsion angle, the distribution of the three other torsion angles is not uniform. Any sampling techniques ignoring this effect would fail to reproduce in this case the appropriate uniform distributions. This bias can be alleviated by properly weighting each configuration of four torsion angles. For an MC approach, the corresponding weight $J(\mathrm{n})$ has to be incorporated in the sampling probability to accept or to refuse a new configuration according to the probability [9]

$$P(\mathrm{m} \to \mathrm{n}) = \min\left\{1, \frac{\alpha(\mathrm{m} \to \mathrm{n}) \cdot \rho(\mathrm{n}) \cdot J(\mathrm{n})}{\alpha(\mathrm{n} \to \mathrm{m}) \cdot \rho(\mathrm{m}) \cdot J(\mathrm{m})}\right\}$$

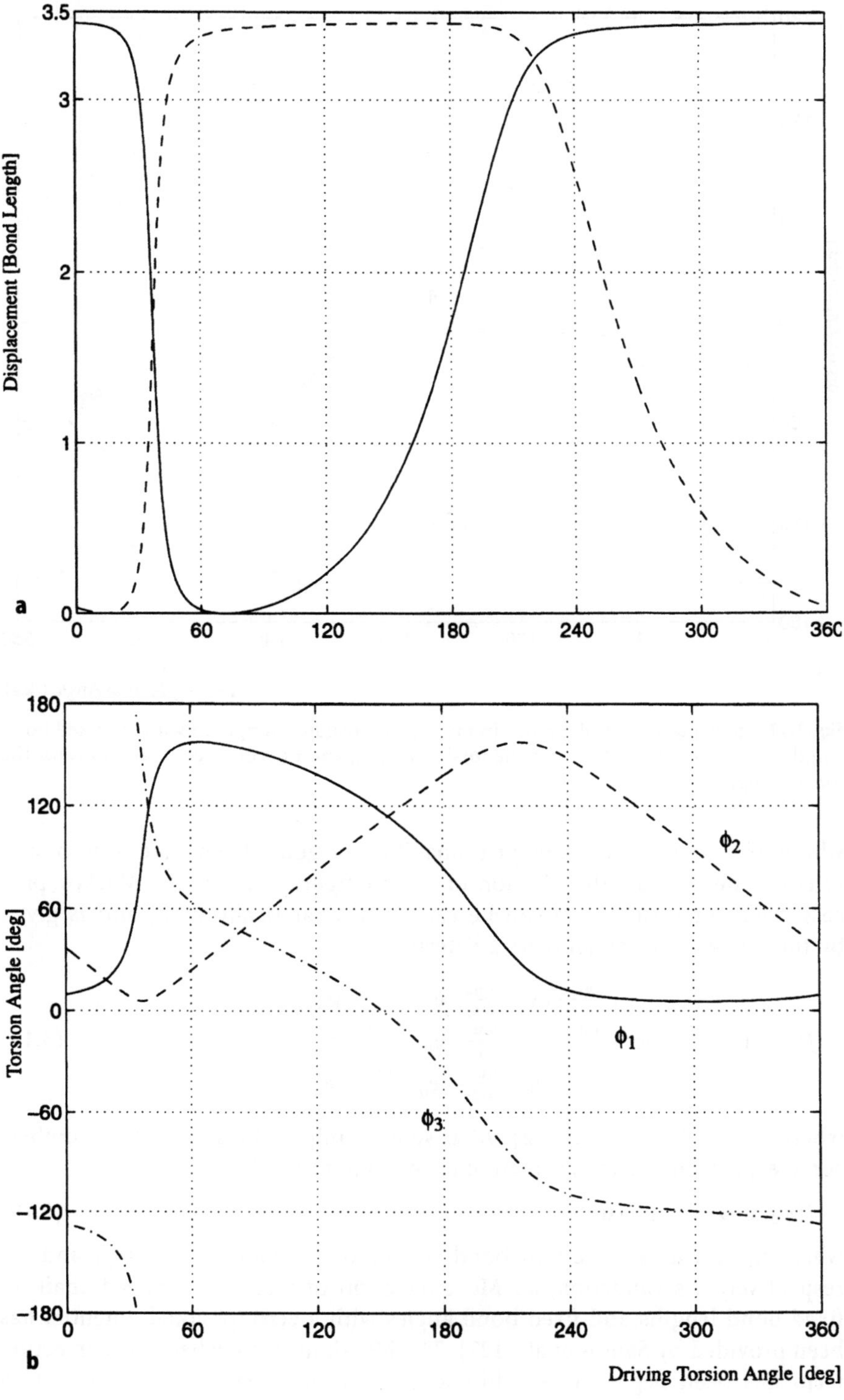
3.5
3
2
1
0
Displacement [Bond Length]
a
0
60
120
180
240
300
360
180
120
60
0
-60
-120
-180
Torsion Angle [deg]
ϕ_2
ϕ_1
ϕ_3
b
Driving Torsion Angle [deg]

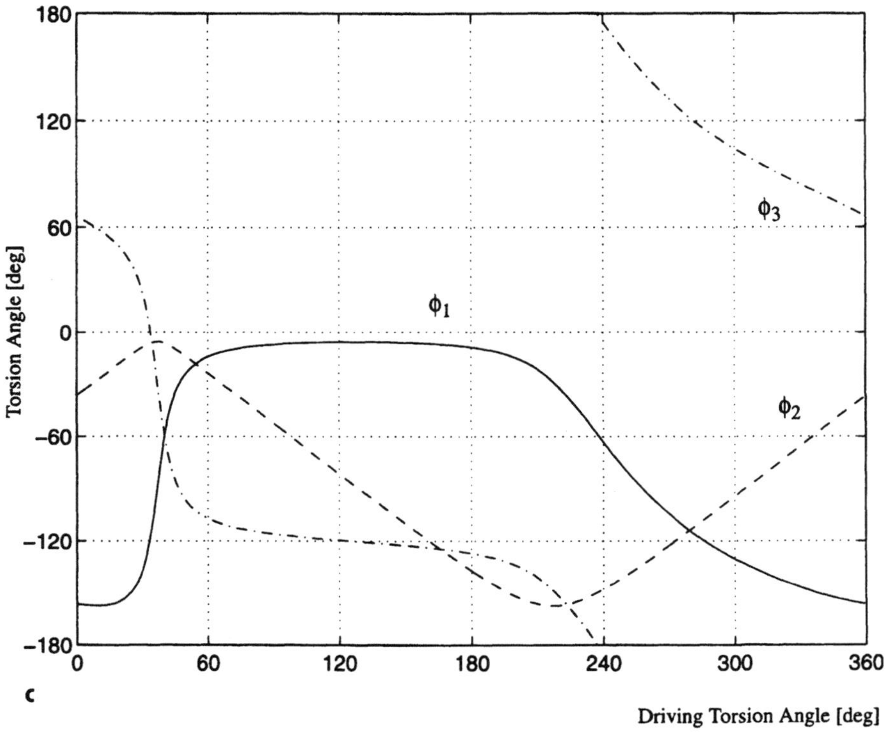

Fig. 3.12. **a** Displacement of a point in the moving residue compared with the fixed bond length. **b** First set of solutions for the torsion angles; and **c** Second set of solutions for the torsion angles

where $\alpha(\mathrm{m} \rightarrow \mathrm{n})$ is the attempt probability for going from state m to n, and $\rho(\mathrm{n})$ the steady-state distribution of the configurations for the Markov process. This correction term can be calculated analytically [127] and is given by the following Jacobian matrix determinant

$$J(\phi_0, \mathbf{u}, \mathbf{u}_\perp) = \det \begin{bmatrix} \frac{\partial \mathbf{u}_\alpha}{\partial \phi_1} \cdot \mathbf{e}_\lambda & \frac{\partial \mathbf{u}_\alpha}{\partial \phi_2} \cdot \mathbf{e}_\lambda & \frac{\partial \mathbf{u}_\alpha}{\partial \phi_3} \cdot \mathbf{e}_\lambda \\ \frac{\partial \mathbf{u}_\beta}{\partial \phi_1} \cdot \mathbf{e}_\nu & \frac{\partial \mathbf{u}_\beta}{\partial \phi_2} \cdot \mathbf{e}_\nu & \frac{\partial \mathbf{u}_\beta}{\partial \phi_3} \cdot \mathbf{e}_\nu \\ \frac{\partial \mathbf{u}_{\perp\gamma}}{\partial \phi_1} \cdot \mathbf{e}_\eta & \frac{\partial \mathbf{u}_{\perp\gamma}}{\partial \phi_2} \cdot \mathbf{e}_\eta & \frac{\partial \mathbf{u}_{\perp\gamma}}{\partial \phi_3} \cdot \mathbf{e}_\eta \end{bmatrix}, \quad (3.18)$$

where $\alpha, \beta, \gamma, \lambda, \mu, \nu \in \{x, y, z\}$. After some manipulations this rather cumbersome expression for the correction term simplifies to [127]

$$J(\phi_0, \mathbf{u}, \mathbf{u}_\perp) = \mathbf{u} \cdot (\mathbf{u}_1 \times \mathbf{u}_2) \quad (3.19)$$

where $\mathbf{u}_1$ and $\mathbf{u}_2$ are the unit bond vectors of the torsion angles ϕ_1 and ϕ_2, respectively. As validation, an MC simulation of a single polybead chain of fixed bond lengths and fixed bond angles with a zero torsional potential has been provided by Santos et al. [127]. The MC simulation, where the correction term has been neglected, exhibits a nonuniform torsion-angle distribution

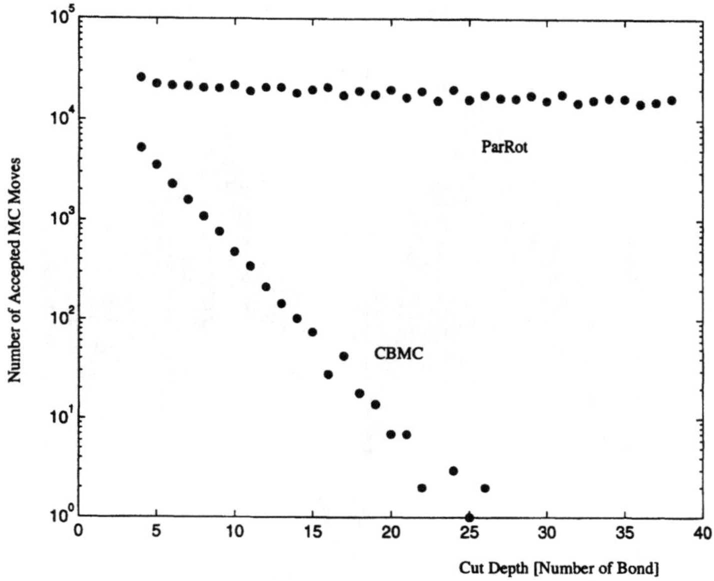

Fig. 3.13. Comparison of the number of successful MC-moves between the CBMC and the ParRot algorithm in dependency of the cut depth for a system of 10 C_{71} at density 0.7 g/cm^3 in a NpT-simulation of 3×10^6 MC steps

(see Fig. 3.14), whereas the correction factor properly compensates for that bias in the distribution.

3.3.4
Results and Discussion

Müller et al. focused on polybead molecules in the united atom approximation as a test system; these are chains formed by spherical methylene beads connected by rigid bonds of length 1.53 Å. The angle between successive bonds of a chain is also fixed at 112°. The torsion angles around the chain backbone are restricted to three rotational isomeric states, the trans (t) and gauche states (g^+ and g^-). The three-fold torsional potential energy function introduced [142] in a study of butane was used to calculate the RIS correlation matrix. "Second order interactions", reflected in the so-called pentane effect, which almost excludes the consecutive combination of g^+g^- states (and vice-versa) are taken into account. In analogy to the polyethylene molecule, a standard RIS-model [143] was used to account for the pentane effect.

Both investigated systems are melts of polybead chains at a temperature of 473 K. Configurations of chain length ranging from 24 beads to 100 beads at a density of 0.75 g/cm^3 (this corresponds to a comparable real system of polyethylene at a pressure of 1 bar) have been successfully embedded. The hard-sphere radius was 2.2 Å, corresponding approximately to a van-der-Waals ra-

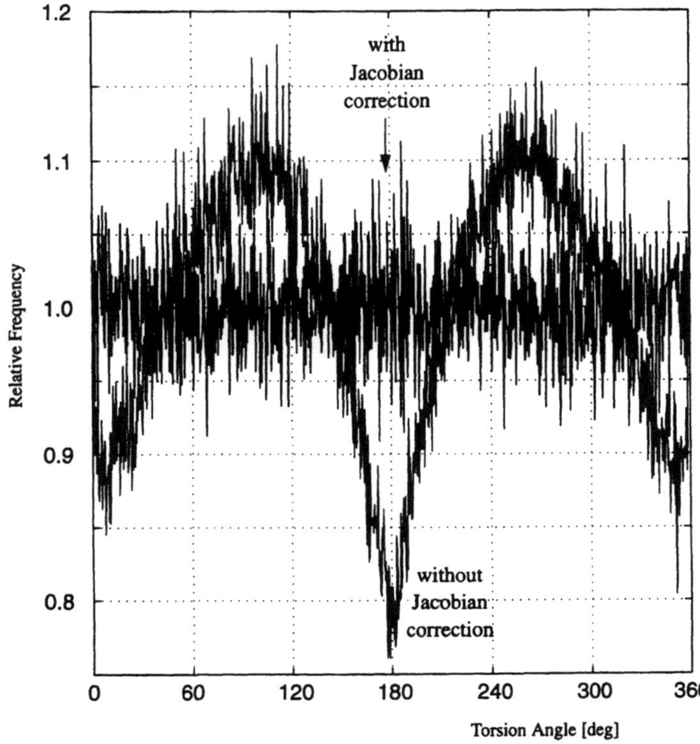

Fig. 3.14. Comparison of the torsion angle density distribution for a MC simulation of an athermal C_{10} phantom chains with Jacobian correction (the horizontal noisy line) and without Jacobian correction (the noisy curve showing two maxima at ca. 100° and 260° and a sharp minimum at 180°)

dius of 3.94 Å from liquid-viscosity comparisons [123]. The RIS-state population at this temperature, calculated by means of the RIS method, is 60.5% of trans. The characteristic ratio, $C_\infty = \lim_{n\to\infty}\langle r^2\rangle/nl^2$, is approximately 7. The statistical probability of encountering g^+g^- pairs is about 1.5%.

The scaling behavior of the embedding algorithm with a varying number of torsion angles has been investigated. The results of runs performed on O2-workstations (MIPS R5000) for 20, 30, and 40 chains are shown in Table 3.1. The required CPU-time obeys a power law of exponent 2.63 ± 0.04 in the number of involved torsion angles, as demonstrated in Fig. 3.15. This is to be compared with the exponent 3, the theoretical value derived from the number of performed optimization loops if the time spent at the optimization stage within the main loop (that aims at gradually lowering the collision limit, and hence, that constitutes the only nondeterministic part of the presented algorithm) is assumed to be independent of the value of the horizon.

The second reported test of the algorithm has been to generate C_{24} polybead chain structures for different numbers of chains: results are shown in

Table 3.1. Generation of dense structures of 20, 30, and 40 polybead chains for different length of chains. The CPU-time refers to the computational time on O_2-workstations (MIPS R5000)

Number of chains	Number of beads per chain	Box edge (Å)	CPU-time (min)	Maximal collision	Minimal encountered distance (Å)
20	26	25.8	54.4	0.116	3.92
	30	27.1	75.6	0.122	3.86
	34	28.3	95.6	0.143	3.77
	38	29.3	143.9	0.150	3.74
	42	30.3	189.5	0.126	3.85
	46	31.3	238.7	0.153	3.73
	50	32.1	286.9	0.163	3.68
	54	33.0	344.6	0.164	3.68
	60	34.1	522.1	0.165	3.68
	68	35.6	691.9	0.176	3.63
	76	36.9	916.7	0.178	3.62
	84	38.2	1170.2	0.176	3.61
	92	39.3	2239.6	0.212	3.47
	100	40.5	2212.7	0.214	3.47
30	26	29.6	107.3	0.123	3.86
	30	31.0	147.1	0.129	3.83
	34	32.3	196.4	0.150	3.74
	38	33.6	355.1	0.135	3.80
	42	34.7	357.9	0.161	3.69
	46	35.8	383.9	0.160	3.70
	50	36.8	564.8	0.154	3.72
	54	37.7	676.7	0.160	3.70
	60	39.0	861.2	0.165	3.67
	68	40.7	1103.4	0.181	3.60
	76	42.2	1777.8	0.180	3.60
	84	43.7	1944.3	0.180	3.56
	92	45.0	2598.6	0.204	3.50
	100	46.3	3245.2	0.213	3.46
40	26	32.5	180.0	0.157	3.71
	30	34.1	201.6	0.157	3.71
	34	35.5	323.1	0.153	3.73
	38	36.9	371.5	0.166	3.67
	42	38.1	638.4	0.164	3.68
	46	39.3	731.2	0.159	3.70
	50	40.4	790.9	0.161	3.69
	54	41.5	1040.0	0.163	3.68
	60	43.0	1714.8	0.174	3.63
	68	44.8	2173.9	0.168	3.66
	76	46.5	3023.9	0.201	3.52
	84	48.0	3131.0	0.212	3.47
	92	49.5	4141.0	0.307	3.05
	100	50.9	4437.9	0.218	3.44

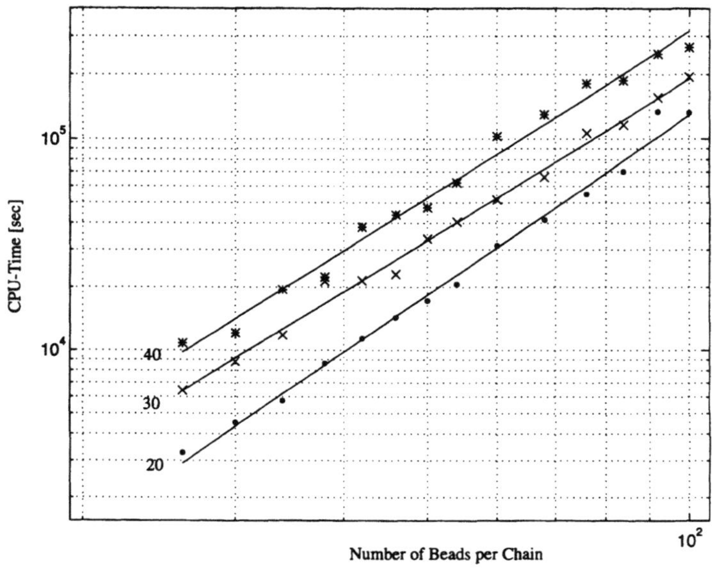

Fig. 3.15. Scaling behavior of the CPU-time for a varying number of united-atoms. The results shown correspond to 20, 30, and 40 chains, respectively

Table 3.2. Generation of dense structures of C_{24} chains for different number of chains. The CPU-time refers to the computational time on DEC 8400-workstations.

Number of chains	Box edge (Å)	CPU-time (min)	Maximal collision	Minimal encountered distance (Å)
20	25.0	35.9	0.209	3.48
30	28.3	72.8	0.160	3.70
59	35.5	234.0	0.210	3.48
69	37.4	299.9	0.213	3.46
120	45.0	891.5	0.217	3.45
140	47.3	1111.3	0.212	3.46
165	50.0	1577.9	0.217	3.45
180	51.4	1980.7	0.216	3.45
200	53.5	2157.8	0.216	3.45
250	57.5	3434.9	0.276	3.19

Table 3.2. The required CPU-time on DEC 8400 workstations obeys a power law of exponent 1.81 ± 0.02 in the number of involved chains as can be seen in Fig. 3.16. In this case, the scaling behavior is less favorable than the theoretical scaling behavior of exponent 1, which is rationalized by Müller et al. [115].

The obtained configurations were not characterized, but when used directly for NpT-MC simulations, which are indeed very sensitive to the "quality" of

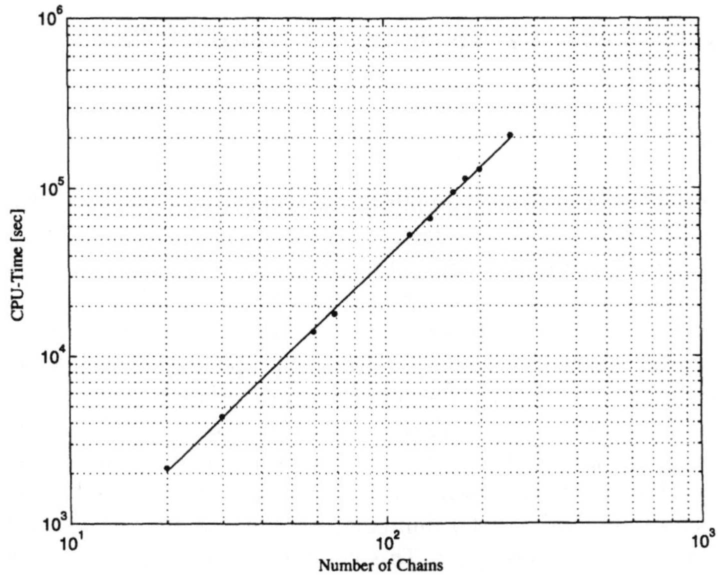

Fig. 3.16. Scaling behavior of the CPU-time for a varying number of chains of C_{24}

the starting configuration, they performed very well. All NpT simulations carried out were stable and Müller et al. concluded that the generated starting configurations are suitable for this kind of application.

3.3.5
Conclusions

These recent results for dense polybead systems are very encouraging. One must wait for tests on realistic polymers with complicated chemical structures and side groups, however, before definitive conclusions can be drawn. The scaling law for the embedding algorithm has to be explored in more detail for the most cumbersome polymer structures.

It appears that the embedding algorithm conjugated with the novel ParRot move constitutes a robust and efficient methodology for generating very "reasonable" initial-guess structures for atomistic simulations. Based on the RIS Ansatz, the embedding algorithm benefits from a great flexibility in the choice of the input parameters that account for the local chain energy configuration; the input for the correlations of torsion angles along the chain backbones can either be calculated with the help of a force field, or extracted from measurements, or even biased in order to study any thinkable structural properties of the macromolecules.

4 Mapping and Reverse Mapping of Coarse-Grained RIS Chains

4.1 Introduction

The discretization on a lattice simplifies the conformational space, because a single occupied site, or bead, represents several atoms from the fully atomistic representation. Thus there are reductions both in the number of elementary particles and degrees of freedom. The number of attempted moves provides a natural time scale for the expression of the dynamics. Simulations can be constructed so that this time scale on the lattice is proportional, in good approximation, to the time scale for a fully atomistic representation of the same system in continuous space [20]. The size of the proportionality constant that relates the two time scales usually gives a computational advantage to the system on the lattice, providing access to slower relaxations than those accessible in conventional molecular dynamics (MD) simulations of the fully atomistic model of the same system in continuous space. This computational advantage is strongest when the simulation can be designed so that the lattice is sparsely occupied.

A major drawback in lattice simulations is the loss of atomistic detail and hence the difficulty in making an unambiguous connection between the chain of beads on the lattice and a specific real polymer of well-defined composition and molecular weight. Most simulations on lattices seek general trends in physical properties that are shared by all flexible chain molecules. When that objective is the only one, the absence of an identification with a specific real polymer is of little concern. Often, however, it is not the general trend, but rather a specific system, with a unique atomic composition, molecular weight distribution, etc., that is of interest. The lattice simulations will be most relevant here if they are designed so that the connection with a specific real system is unambiguous. Ideally, a fully atomistic replica in continuous space should be easily recoverable from any replica of the coarse-grained system in the discrete space on the lattice. With this objective in mind, a group at Akron is developing techniques that have the attributes described below for the three numbered steps depicted by solid arrows in Fig. 4.1 [144].

Step 1. Mapping: A fully atomistic model of a specific amorphous polymer in continuous space is mapped at bulk density onto a coarse-grained representation on an appropriately chosen high coordination lattice. The lattice is of high coordination so that it can accomodate a variety of local conformations. The specific high coordination lattice is selected so that this set of local conformations includes suitable approximations to the preferred local conformations of the real polymer of interest. The mapping produces a coarse-grained representation of the real chain, thereby reducing the number of fundamental particles so that the subsequent simulation on the lattice will be computationally efficient, even when the system is at bulk density.

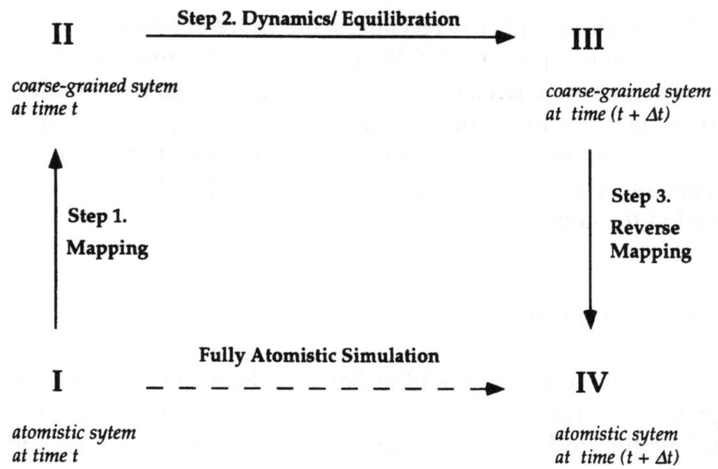

Fig. 4.1. Schematic representation of three numbered steps in a MC simulation on a high coordination lattice (*solid arrows*) that replace a simulation of the fully atomistic system in continuous space (*single dashed line*)

Step 2. Dynamics/Equilibration: The system evolves through time by simulation on the high coordination lattice. The Hamiltonian that is used in the usual Metropolis criteria [105] for the acceptance of a proposed move must have a local, intrachain component that is system specific, so that the coarse-grained chains on the high coordination lattice will retain the local conformational characteristics of the real atomistic chains from which they were derived in *Step 1*. The Hamiltonian must contain another term, also system specific, that governs the interchain interactions and long-range intrachain interactions in a manner that produces reasonable densities and cohesive energies.

Step 3. Reverse Mapping: The final step recovers a fully atomistic description of the system in continuous space from arbitrarily chosen points in the trajectory. For this "reverse mapping" to be successful, a robust methodology must be employed for that portion of the Hamiltonian in *Step 2* that governs the local intrachain interactions. The reverse mapping must restore chain atoms (and bonds) that were temporarily removed from the description of the system during the mapping in *Step 1*. Even though these atoms were absent during *Step 2*, the Hamiltonian must take their interactions into account in describing the coarse-grained bonds that are retained, if these missing atoms are to be successfully restored to the system in a physically acceptable manner at arbitrarily chosen points along the trajectory.

Recent and current work along these lines, most of which uses monodisperse *n*-alkane melts as the systems of interest, is reviewed here. The implementation of *Steps 1–3* for this class of molecules is described first. Then the results obtained by this method for unentangled n-alkane melts are briefly compared with those reported recently using more conventional MD simulations in continuous space with fully atomistic [145] and united atom [146] representations of the chain. The computational efficiency of the dynamics in *Step 2* is compared with a conventional MD simulation of the fully atomistic systems, as implemented in a popular commercial software package. A few other types of systems susceptible to simulation by this technique will be mentioned at the conclusion.

4.2
The First Step – Mapping

The choice of the high coordination lattice is based on the local conformational preferences for the chain of interest. The validity of this assertion is illustrated by the work of Skolnick and co-workers on the folding and unfolding of specific globular proteins [147–150]. The motivation for simulation of these processes on lattices was supplied by regularities in the distances between successive C^α atoms, strong preferences for certain ranges of the angles defined by three consecutive C^α atoms, and preferences for certain ranges of the torsions defined by four consecutive C^α atoms [150]. When a diamond lattice is used for this purpose, some features, such as the Greek-key, β-barrel, could be approximated [148–149], but this lattice could not capture other important local features of protein structure, such as the packing of α helices and β sheets in a parallel orientation, and allowing helices with 3 2/3 (rather than 4) monomer units per turn. A more accurate mapping of the preferred local conformations of proteins was achieved on the high coordination (24 nearest neighbors) knight-move lattice, which is derived from an underlying cubic lattice and employs [2,1,0] basis vectors that join successive C^α atoms [147]. The C^α atoms of the chain describe a knight's walk in three dimensions. The work on globular proteins highlights the importance of selecting a high coordination lattice based on an appropriate underlying lattice, with judicious choice of the basis vectors.

The rotational isomeric state (RIS) model is frequently employed for the analysis of conformation-dependent physical properties of unperturbed chains, such as the mean square unperturbed radius of gyration, $\langle R_G^2 \rangle_0$ [35–36]. RIS models rarely describe a lattice, but they do provide a convenient description of the preferred conformations of a real chain in terms of discrete values for the torsion angles (ϕ), bond angles (θ), and bond lengths (ℓ). Several RIS models have been proposed for unperturbed polyethylene (PE) chains [124]. The most widely used RIS model for PE, described initially by Abe et al. [143], uses a single bond length ℓ_{CC} = 1.54 Å, a single bond angle $\theta = 112°$, and three torsion angles ϕ of 180° and $\pm(60° + \Delta\phi)$ for trans (t) and gauche$^\pm$ ($g^\pm$) states. The conformational energy surfaces reported by Abe et al. imply that the value of $\Delta\phi$ is in the range 0–10°. This RIS model fits exactly onto a diamond lattice when $\Delta\phi = 0°$ and the bond angle is lowered from 112° to the tetrahedral angle. Therefore one suspects that a PE chain might be more successfully mapped onto a high coordination lattice based on an underlying diamond lattice, instead of an underlying cubic lattice, in contrast to the polypeptide chains studied by Skolnick and co-workers [147–150], for which the opposite result was obtained. Investigation of numerous high coordination lattices based on these underlying lattices verifies the truth of that suspicion [151]. The contrasting results for PE and polypeptides show that the type of underlying lattice for the mapping procedure should be chosen with an eye on the preferred local conformations adopted by the chain of interest.

If the diamond lattice itself is used for the mapping of the PE chains, each internal bead represents a methylene unit, the step length is the C–C bond length, and C_xH_{2x+2} is represented by x beads. Typical bulk densities for n-alkane melts, which are in the range 0.7–0.8 g/cm^3, are achieved with occupancy of 16–19 % of the sites on this lattice.

The number of occupied sites can be reduced by a factor of two if the mapping procedure discards every second carbon atom in the chains. C–C bonds no longer appear discretely in the model of the chain. This change in the lattice is depicted in Fig. 4.2.A and 4.2.B. The light and dark sites in Fig. 4.2.A describe a diamond lattice, where the light sites alternate with the dark sites. Only the dark sites are retained in the lattice depicted in Fig. 4.2.B. The lattice depicted in Fig. 4.2.B is well known. It is identical to the fcc (face centered

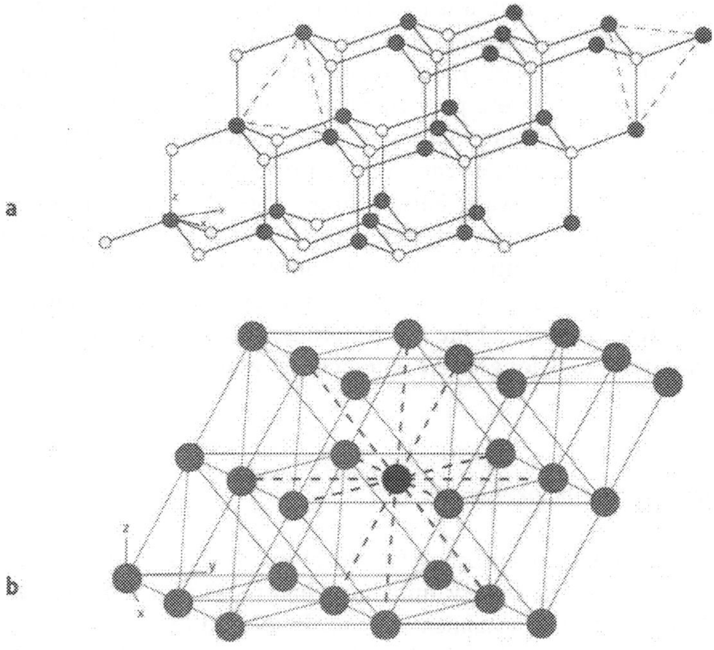

Fig. 4.2. **A** Diamond lattice with sites alternately represented by open and shaded spheres. The two dashed equilateral triangles are discussed in the text and **B** The 2nnd lattice obtained by rejection of the open spheres from **A**

cubic) closest packing of uniform hard spheres. Our term for the lattice in Fig. 4.2.B is the "second nearest neighbor diamond lattice", or 2nnd lattice [151], because this name conveys (in a way that "uniform packing of hard spheres" does not) a crucial relationship to an underlying diamond lattice. The number of sites in shell k is $z_k = 10\ k^2 + 2$ [152]. The unit cell can be represented by a distorted cube, and the sites on the 2nnd lattice are easily identified with integers. If the step length on the underlying diamond lattice is ℓ, the step length on the 2nnd lattice, L, is larger by a factor of $2\sqrt{2/3}$. For the simulation of a PE melt, $\ell_{CC} = 1.53$ and L = 2.50 Å. The occupancy of both lattices is the same for PE; each chain occupies half as many beads on the 2nnd lattice as on the diamond lattice, but the 2nnd lattice contains only half as many sites as the diamond lattice. The occupancy of the 2nnd lattice is lower (near 12%) in simulations of polypropylene melts, where each site on the 2nnd lattice represents a C_3H_6 unit.

4.3
The Second Step – MC Simulation

The conformation and dynamics of simple isolated model chains have been investigated on the 2nnd lattice [151]. In these simulations, a bead may move from its present position to a randomly chosen nearest neighbor site on the 2nnd lattice, provided the move preserves its distance from the two beads (one bead for terminal beads) to which it is connected. These single bead moves on the 2nnd lattice correspond to three- or four-bond crankshaft motions of the fully atomistic chain on the underlying diamond lattice [144]. Simulations using the single-bead moves were performed for isolated chains that behave as random walks, non-reversal random walks, and self-avoiding random walks [151]. The chain on the 2nnd lattice cannot pass through itself when it is treated as a self-avoiding random walk. For all three types of random walks, the mean square end-to-end distance, relaxation time for the end-to-end vector, and translational diffusion coefficient for the center of mass scaled with the number of beads in the manner expected from theory. The present purpose now requires two modifications of the self-avoiding random walk: short-range intrachain interactions and long-range interactions appropriate for the real chain.

4.3.1
Short-Range Intramolecular Interactions

The local conformational preferences of a PE chain are described by more complicated torsion potential energy functions than those in a random walk. The simulation must not only establish the coordinates on the 2nnd lattice of every second carbon atom in the initial configurations of the PE chains, but must also describe the intramolecular short range interactions of these carbon atoms, as well as the contributions to the short-range interactions from that half of the carbon atoms that are not represented discretely on the 2nnd lattice. The RIS model can be adapted to this purpose.

In the RIS model for PE described by Abe et al. [143], the conformational partition function, Z, of the unperturbed chain with n bonds is formulated using a well-known statistical weight matrix for bond i, U_i, that is usually written in a 3 × 3 form.

$$Z = \begin{bmatrix} 1 & 0 & 0 \end{bmatrix} \begin{bmatrix} 1 & \sigma & \sigma \\ 1 & \sigma & \sigma\omega \\ 1 & \sigma\omega & \sigma \end{bmatrix}^{n-2} \begin{bmatrix} 1 \\ 1 \\ 1 \end{bmatrix} \tag{4.1}$$

The columns are indexed by the three states of bond i, the rows are indexed by the three states of bond i–1, and the order of indexing is t, g^+, g^-. The statistical weight denoted by σ, which is approximately exp (–250/T), is the first-order weight of a gauche state relative to a trans state. The second-order interaction, which arises from the pentane effect where a bond pair is $g^{\pm}g^{\mp}$, is weighted by ω, which is approximately exp (–1000/T).

In the RIS model for use on the 2nnd lattice, a step of length L on the 2nnd lattice incorporates two C–C bonds of length ℓ_{CC} from the fully atomistic representation. The two consecutive bonds of length ℓ_{CC} on the diamond lattice always form a tetrahedral bond angle, but two consecutive bonds of length L on the 2nnd lattice can form a bond angle of 60°, 90°, 120°, or 180°. (Immediate reversals, for which the bond angle is 0°, are disallowed because self-intersection is prohibited.) Whereas the state of a C–C bond on the underlying diamond lattice (and in the conventional RIS model) is determined by a single torsion angle, the state of a bond of length L on the 2nnd lattice is specified by the states of two consecutive C–C bonds of length ℓ_{CC} in the atomistic chain, as shown in Fig. 4.3. The torsion angles at the two C–C bonds that meet at bead i determine the angle between two successive steps (bead i–1 to i, bead i to i+1) in the coarse-grained chain on the 2nnd lattice. Thus the states of four consecutive C–C bonds must be considered in the construction of the statistical weight matrix that will be employed for an internal bond of length

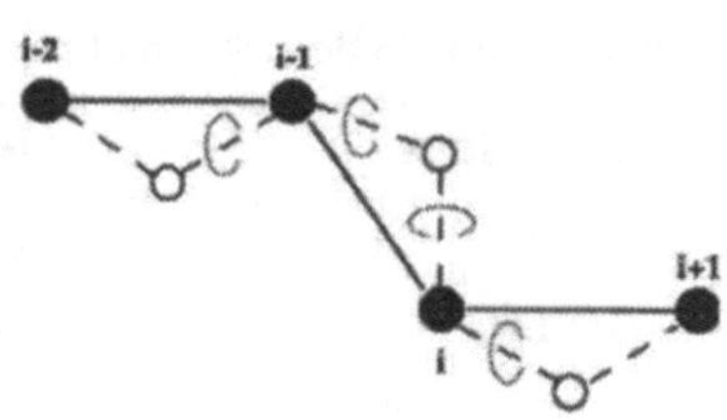

Fig. 4.3. Torsions angles at bonds of length ℓ that change the angle between a pair of steps of length L. The two torsions on the left determine the angle i–2, i–1, i, and the two torsions on the right determine the angle i–1, i, i+1

L on the 2nnd lattice. The rows and columns are indexed in the order tt, tg^+, tg^-, g^+t, g^-t, g^+g^+, g^-g^-, g^+g^-, g^+g^-. Then the matrix is [153]

$$\mathbf{U} = \begin{bmatrix} 1 & \sigma & \sigma & \sigma & \sigma & \sigma^2 & \sigma^2 & \sigma^2\omega & \sigma^2\omega \\ \\ 1 & \sigma & \sigma & \sigma & \sigma\omega & \sigma^2 & \sigma^2\omega & \sigma^2\omega & \sigma^2\omega^2 \\ 1 & \sigma & \sigma & \sigma\omega & \sigma & \sigma^2\omega & \sigma^2 & \sigma^2\omega^2 & \sigma^2\omega \\ 1 & \sigma & \sigma & \sigma & \sigma & \sigma^2 & \sigma^2 & \sigma^2\omega & \sigma^2\omega \\ 1 & \sigma & \sigma & \sigma & \sigma & \sigma^2 & \sigma^2 & \sigma^2\omega & \sigma^2\omega \\ \\ 1 & \sigma & \sigma & \sigma & \sigma\omega & \sigma^2 & \sigma^2\omega & \sigma^2\omega & \sigma^2\omega^2 \\ 1 & \sigma & \sigma & \sigma\omega & \sigma & \sigma^2\omega & \sigma^2 & \sigma^2\omega^2 & \sigma^2\omega \\ \\ 1 & \sigma & \sigma & \sigma\omega & \sigma & \sigma^2\omega & \sigma^2 & \sigma^2\omega^2 & \sigma^2\omega \\ 1 & \sigma & \sigma & \sigma & \sigma\omega & \sigma^2 & \sigma^2\omega & \sigma^2\omega & \sigma^2\omega^2 \end{bmatrix} \tag{4.2}$$

The motivation for writing U with 4^2 blocks will become apparent later.

For a chain constructed from bonds subject to a symmetric torsion potential energy function, as is the case for PE, the dimensions of this matrix can be reduced by focusing on the distance between beads i–1 and i+1, and introducing coarse-grained statistical weights. The basis for this procedure becomes apparent upon consideration of the conformations of *n*-pentane. In the RIS model of Abe et al. [143], there are four end-to-end distances for *n*-pentane, which, on a diamond lattice with step length ℓ_{CC} have the values shown in Table 4.1. Three successive beads, indexed i–1, i, and i+1, on the 2nnd lattice, correspond to the terminal methyl groups and the middle methylene group in *n*-pentane. The distance between i–1 and i, and between i and i+1, must be L. However, the distances between i–1 and i+1 can have any of the four values listed in Table 4.1, depending on the states at the two internal C–C bonds of length ℓ_{CC}. If the end-to-end distance is 2L, both internal C–C bonds must be in t states. For the three other distances, however, an ambiguity arises. For example, if the distance between beads i–1 and i+1 on the 2nnd lattice is $\sqrt{3}$ L, one of the internal C–C bonds must be in a t state and the other in a g state, but we cannot distinguish which one is gauche, nor whether it is g^+ or g^-,

Table 4.1. End-to-end distances in the RIS description of n-pentane[a]

Conformation	2nnd lattice	Diamond lattice
tt	2L	$4\sqrt{2/3}\ell_{CC}$
$tg^{\pm}$, $g^{\pm}t$	$\sqrt{3}$ L	$2\sqrt{2}\ell_{CC}$
$g^{\pm}g^{\pm}$	$\sqrt{2}$ L	$4\sqrt{1/3}\ell_{CC}$
$g^{\pm}g^{\mp}$	L	$2\sqrt{2/3}\ell_{CC}$

[a] The step lengths are L on the 2nnd lattice and ℓ_{CC} on the underlying diamond lattice.

because the C–C bonds themselves are missing on the 2nnd lattice. This ambiguity can be resolved in two ways, described in the next two sections.

4.3.1.1
Next-Nearest Neighbor Distances

A simple way of treating the local intrachain structure on the 2nnd lattice is to focus on the distance between next-nearest neighbor beads. In this approximation, the 2nnd lattice can only accept information from a 9×9 statistical matrix that is of the form [153]

$$\mathbf{U}_{2\mathrm{nnd}} = \begin{bmatrix} 1 & \sigma & \sigma & \sigma & \sigma & \sigma^2 & \sigma^2 & \sigma^2\omega & \sigma^2\omega \\ 1 & a & a & a & a & \sigma b & \sigma b & \sigma\omega b & \sigma\omega b \\ 1 & a & a & a & a & \sigma b & \sigma b & \sigma\omega b & \sigma\omega b \\ 1 & a & a & a & a & \sigma b & \sigma b & \sigma\omega b & \sigma\omega b \\ 1 & a & a & a & a & \sigma b & \sigma b & \sigma\omega b & \sigma\omega b \\ 1 & b & b & b & b & c & c & \omega c & \omega c \\ 1 & b & b & b & b & c & c & \omega c & \omega c \\ 1 & b & b & b & b & c & c & \omega c & \omega c \\ 1 & b & b & b & b & c & c & \omega c & \omega c \end{bmatrix} \tag{4.3}$$

where, by the way the 9×9 matrix is organized into 4^2 blocks, we recognize that there are only four distinguishable states (based on the separation of beads i–1 and i+1) for bond i on the 2nnd lattice, as was shown in Table 4.1. This approximation is most likely to be useful if the bonds in the chain are subject to a symmetric torsion potential energy function. It would not be suitable for a description of vinyl polymers, because the weighting scheme presented in Eq. (4.3) could not draw distinctions between chains of different stereochemical compositions and stereochemical sequences. The correct treatment of chains in which the bonds are subject to asymmetric torsion potential energy functions, as is the case in vinyl polymers, will be presented in Sec. 4.3.1.2.

The mapping of the conventional RIS model for PE onto its coarse-grained equivalent on the 2nnd lattice, using the approximation presented in Eq. (4.3), introduces several coarse-grained statistical weights, which are denoted by a, b, and c in this equation. Comparison of equivalent blocks in Eqs. (4.2) and (4.3) raises questions about the values of these coarse-grained statistical weights. For example, "a" must have a single value in Eq. (4.3), but comparison of the 4×4 blocks in Eq. (4.2) and (4.3) shows that the conventional RIS model would usually assign "a" as σ, but sometimes it becomes $\sigma\omega$ instead. The ability to distinguish between these two assignments is lost when the short-range interactions are described in terms of the distances between next-nearest neighbor beads on the 2nnd lattice. There are a variety of methods for providing acceptable values of the coarse-grained statistical weights [153–155]. A simple assignment as the geometric mean of the statistical weights in the appropriate block of Eq. (4.2) ($a = \sigma\omega^{1/8}$, $b = \sigma\omega^{1/4}$, $c = \sigma^2\omega^{1/2}$) works surprisingly well for PE [153], but fails for polyoxyethylene [154]. A general method demands that identical values must be obtained for the mean square unperturbed end-to-end distance, $\langle r^2 \rangle_0$, in the limit of a large number of carbon atoms, for the chain of n C–C bonds of length ℓ_{CC} described by Eq. (4.1) and the chain of n/2 bonds of length L described by Eq. (4.3). This requirement is easily satisfied [155].

The identity of several rows and columns in the matrix in Eq. (4.3) shows that the same information is contained in a more compact matrix. This matrix can be written with the four distinguishable states on the 2nnd lattice indexing the rows and columns, in the order in which they are listed in Table 4.1 [153]:

$$\mathbf{U}_{2nnd} = \begin{bmatrix} 1 & 4\sigma & 2\sigma^2 & 2\sigma^2\omega \\ 1 & 4a & 2\sigma b & 2\sigma\omega b \\ 1 & 4b & 2c & 2\omega c \\ 1 & 4b & 2c & 2\omega c \end{bmatrix} \tag{4.4}$$

or in a still more condensed form,

$$\mathbf{U}_{2nnd} = \begin{bmatrix} 1 & 4\sigma & 2\sigma^2(1+\omega) \\ 1 & 4a & 2\sigma b(1+\omega) \\ 1 & 4b & 2c(1+\omega) \end{bmatrix} \tag{4.5}$$

General rules for the formulation of these matrices for other polymers are presented in an Appendix to Rapold and Mattice [153].

The probability for a conformation of an unperturbed chain of n C–C bonds on the underlying diamond lattice can be expressed in terms of the probability for state η at the second bond and the conditional probability for state η at bond i ($2 < i < n$), given state ξ at the preceding bond. These probabilities, which are usually denoted by $p_{\eta;2}$ and $q_{\xi\eta;i}$, respectively, are extracted from Z of Eq. (4.1) using the conventional partial derivatives [35,36]. Then the probability, p, for a specified conformation of the entire chain is:

$$p = p_{\alpha;2}q_{\alpha\beta;3}q_{\beta\gamma;4}\cdots q_{\psi\omega;n-1} \tag{4.6}$$

A perfectly analogous procedure can be performed for the RIS model on the 2nnd lattice, using an expression for Z developed with the statistical weight matrix in Eq. (4.4) or (4.5) [153]. The replacement for Eq. (4.6) becomes:

$$p = P_{\alpha;1}Q_{\alpha\beta;2}Q_{\beta\gamma;3}\cdots Q_{\psi\omega;(n/2)-1} \tag{4.7}$$

where upper case has been used for the probabilities derived from the coarse-grained RIS model for the chain on the 2nnd lattice, and the subscripts $\alpha\ldots\omega$ now refer to the states of the bonds of length L on this lattice.

If bead i on the 2nnd lattice moves to a neighboring site, there are changes in the vector that connects beads i-2 and bead i, as well as the vector that connects beads i and i+2. Provided that bead i is at least four beads from the nearest end of the chain, the ratio of the probabilities of the new and old conformation, via application of Eq. (4.7) to both conformations, is [153]:

$$\left(\frac{p_{new}}{p_{old}}\right)_{local} = \frac{Q_{\epsilon\zeta^*;i-2}Q_{\zeta^*\eta;i-1}Q_{\eta\iota^*;i}Q_{\iota^*\kappa;i+1}}{Q_{\epsilon\zeta;i-2}Q_{\zeta\eta;i-1}Q_{\eta\iota;i}Q_{\iota\kappa;i+1}} \tag{4.8}$$

where the asterisks denote the states that have changed in the switch from the old to the new conformation, due to the change in location of bead i. Equations for this ratio for the special cases where the moving bead is close to a chain end, which are constructed in an analogous manner, have been given by Rapold and Mattice [153]. The ratio $(p_{new}/p_{old})_{local}$ can be used directly in the Metropolis criterion for accepting the move [105]. If the only interactions employed in the simulations are the local interactions:

$$p_{move} = \min\left[1, \left(\frac{p_{new}}{p_{old}}\right)_{local}\right] \tag{4.9}$$

Using this procedure, the classic RIS model for the atomistic chain has been adapted so that it influences the conformations populated by the coarse-grained chain on the high coordination lattice.

4.3.1.2
Orientation of Two Successive Bond Vectors of Length L

In the previous section, the adaptation of the RIS model was based on the distance between next-nearest neighbor beads. This approach is obviously inadequate for CH_3-CHX-CH_2-CHX-CH_3, because it necessarily abandons the ability to attribute different conformational characteristics to the meso and racemo stereoisomers. Therefore a more robust adaption of the RIS model to the 2nnd lattice is necessary if one wants to investigate the influence of stereochemical composition and stereochemical sequence on vinyl polymers [156]. Here we describe a method that has this capability. Of course, this method retains the ability to treat chains such as PE in which the bonds are subject to symmetric torsion potential energy functions.

The 2nnd lattice depicted in Fig. 4.2.B can be converted to a diamond lattice by inserting a bond of length ℓ upward from each 2nnd bead. By so doing, we recover the open spheres in Fig. 4.2.A. Alternatively, bonds of length ℓ could have been inserted in the opposite direction (downward from the filled circles in Fig. 4.2.B), thereby establishing a different set of positions for the open spheres. Both operations recover a diamond lattice characterized by the same step length ℓ, but the two diamond lattices share only half of their sites. In the present method, we must decide at the beginning of the simulation which of these two diamond lattices will be associated with the 2nnd lattice. This requirement implies that it would be misleading to say that the simulation is performed on a lattice identical to the closest packing of hard spheres, because such a statement does not imply a necessary relationship with one of the two nonidentical underlying diamond lattices.

Table 4.2. Coordinates of the twelve nearest neighbors for bead j (located at x_j, y_j, z_j), in terms of the planes defined by the distorted cube in Fig. 4.2.B

x	y	z
x_j	y_{j+1}	z_j
x_j	y_{j-1}	z_j
x_{j+1}	y_j	z_j
x_{j-1}	y_j	z_j
x_{j+1}	y_{j-1}	z_j
x_{j-1}	y_{j+1}	z_j
x_j	y_j	z_{j+1}
x_j	y_j	z_{j-1}
x_j	y_{j-1}	z_{j+1}
x_j	y_{j+1}	z_{j-1}
x_{j-1}	y_j	z_{j+1}
x_{j+1}	y_j	z_{j-1}

Equivalent simulations can be performed with either choice of the relationship between the 2nnd lattice and its underlying diamond lattice. With whichever choice is made, there are twelve possible vectors from a bead on the 2nnd lattice to its nearest neighbors. The components of these vectors are specified in Table 4.2 in units of the step length along the three axes of the 2nnd lattice. As a compact notation for these vectors, we will use + for +1, and - for -1, in units of the spacing (L) on the 2nnd lattice, so that $+\ 0\ -$ is the abbreviation for the vector

$$\begin{bmatrix} +1 \\ 0 \\ -1 \end{bmatrix} \tag{4.10}$$

Consider two such vectors, the first from bead i-1 to i, and the second from bead i to i+1. Self-avoidance disallows combinations where the angle between the two vectors is 180°, so that they point in opposite directions, because beads i$\pm$1 would then occupy the same site on the 2nnd lattice. At the other extreme, where the angle between these two vectors is 0°, so that they point in the same direction, the chain on the underlying diamond lattice must have trans states at the two bonds of length ℓ that meet at bead i. For all other combinations of the two vectors, the states at these two bonds of length ℓ cannot be specified until we select which of the two distinct diamond lattices underlies the 2nnd lattice. With one choice of the diamond lattice, the conformations of these two bonds are given by the entries in the 12×12 array in Table 4.3 [156]. Diagonal lists of tt and g^+g^+ run from the upper left to the lower right, and diagonal lists of "rev" (for the prohibited immediate rever-

Table 4.3. Conformations at the two bonds of length ℓ to bead i in terms of the vectors i-1 → i (rows) and i → i+1 (columns) for bond vectors of length L for one relationship between the 2nnd lattice and its underlying diamond lattice

	00+	+00	0+0	+0–	0–+	–+0	+–0	0+–	–0+	0–0	–00	00–
00+	tt	g^+t	g^-t	g^+g^-	tg^-	g^-g^-	g^+g^+	g^-g^+	tg^+	col	col	rev
+00	g^-t	tt	g^+t	tg^-	g^-g^-	g^+g^-	tg^+	g^+g^+	g^-g^+	col	rev	col
0+0	g^+t	g^-t	tt	g^-g^-	g^+g^-	tg^-	g^-g^+	tg^+	g^+g^+	rev	col	col
+0–	col	tg^+	g^-g^-	tt	col	g^-g^+	tg^-	g^-t	rev	g^+g^+	g^+g^-	g^+t
0–+	tg^+	g^-g^-	col	g^-g^+	tt	col	g^-t	rev	tg^-	g^+t	g^+g^+	g^+g^-
–+0	g^-g^-	col	tg^+	col	g^-g^+	tt	rev	tg^-	g^-t	g^+g^-	g^+t	g^+g^+
+-0	g^+g^+	tg^-	col	tg^+	g^+t	rev	tt	col	g^+g^-	g^-t	g^-g^+	g^-g^-
0+–	col	g^+g^+	tg^-	g^+t	rev	tg^+	g^+g^-	tt	col	g^-g^+	g^-g^-	g^-t
–0+	tg^-	col	g^+g^+	rev	tg^+	g^+t	col	g^+g^-	tt	g^-g^-	g^-t	g^-g^+
0–0	g^-g^+	g^+g^-	rev	g^+g^+	g^-t	col	g^+t	col	g^-g^-	tt	tg^+	tg^-
–00	g^+g^-	rev	g^-g^+	col	g^+g^+	g^-t	col	g^-g^-	g^+t	tg^-	tt	tg^+
00–	rev	g^-g^+	g^+g^-	g^-t	col	g^+g^+	g^-g^-	g^+t	col	tg^+	tg^-	tt

sals) and g^-g^- run from the lower left to the upper right. The 12×12 matrix can be constructed from sixteen 3×3 blocks.

$$\begin{bmatrix} \mathbf{A} & \mathbf{B} & \mathbf{A} & \mathbf{B} \\ \mathbf{B} & \mathbf{A} & \mathbf{B} & \mathbf{A} \\ \mathbf{A} & \mathbf{B} & \mathbf{A} & \mathbf{B} \\ \mathbf{B} & \mathbf{A} & \mathbf{B} & \mathbf{A} \end{bmatrix} \tag{4.11}$$

where the blocks are either

$$\mathbf{A} = \begin{bmatrix} \mathrm{a} & \mathrm{b} & \mathrm{c} \\ \mathrm{c} & \mathrm{a} & \mathrm{b} \\ \mathrm{b} & \mathrm{c} & \mathrm{a} \end{bmatrix} \tag{4.12}$$

where “a” is tt or g^+g^+, or

$$\mathbf{B} = \begin{bmatrix} \mathrm{b} & \mathrm{c} & \mathrm{d} \\ \mathrm{c} & \mathrm{d} & \mathrm{b} \\ \mathrm{d} & \mathrm{b} & \mathrm{c} \end{bmatrix} \tag{4.13}$$

where d is rev or g^-g^-. In all sixteen blocks, b and c are chosen from tg^+, tg^-, g^+t, g^-t, g^+g^-, g^-g^+, and “col”, where col denotes an unphysical local collapse, which will be described later. In one of the 3×3 blocks, b = c = col, but b and c are distinct in the remaining fifteen blocks.

For the other choice of the relationship between the 2nnd lattice and its underlying diamond lattice, the entries in Table 4.3 must be modified according to the rules in Table 4.4. The only ambiguity in Table 4.4 is whether a specific local collapse, “col”, is to be replaced with g^+g^- or g^-g^+. This ambiguity can easily be resolved [156].

Table 4.4. Replacements for the C-C bonds pairs to a bead on the 2nnd lattice upon changes in the relationship between this lattice and its underlying diamond lattice.

col	$\rightleftharpoons$	g^+g^- or g^-g^+
g^+g^+	$\rightleftharpoons$	g^-g^-
rev	$\rightleftharpoons$	rev (no change)
tg^+	$\rightleftharpoons$	g^+t
tg^-	$\rightleftharpoons$	g^-t
tt	$\rightleftharpoons$	tt (no change)

The term "collapse" refers to a physically untenable situation in which the regeneration of the atomistic representation on the diamond lattice from the coarse-grained chain on the 2nnd lattice assigns two new carbon atoms to the same site on the diamond lattice, so that the two new carbon atoms "collapse" into one [144]. Table 4.4 shows that the local collapse is somehow related (on the 2nnd lattice) to the physically tenable g^+g^- or g^-g^+ combinations, and Table 4.3 (in conjunction with Table 4.4) shows that exactly 12 combinations of vectors specify a local collapse, while another twelve combinations specify the tenable g^+g^- and g^-g^+. In all 24 of these cases, three bonded beads on the 2nnd lattice describe an equilateral triangle of area $L^2/2$. If these three beads are the three shaded circles connected by the dashed lines to the left in Fig. 4.2.A, the open circles of the diamond lattice can be inserted as expected for a $g^\pm g^\mp$ conformation, as shown in that figure. However this same diamond lattice has a single site located at the center of mass of the three shaded circles on the right in Fig. 4.2.A. We denote this latter situation as a local "collapse".

Prohibitation of reversals can be built into the simulation on the 2nnd lattice without having to specify its relationship to the underlying diamond lattice. We simply disallow all moves that cause two successive vectors to point in opposite directions. But in order to also disallow the unphysical local collapses, we must know which of the equilateral triangles of area $L^2/2$ correspond to local collapses, and which correspond to tenable $g^\pm g^\mp$ pairs. This sorting of the equilateral triangles is possible if, and only if, we have defined which diamond lattice underlies the 2nnd lattice.

When the simulation is based on the orientations of pairs of vectors, examination of all such pairs for a chain on the 2nnd lattice immediately specifies the states of all of the bonds of length ℓ on the underlying diamond lattice, as is apparent from Table 4.3. If the chain were PE, its statistical weight could be assigned using the RIS model described by Abe et al. [143]. Alternatively, if the chain were polypropylene (PP) with a specified stereochemical sequence, its statistical weight could be assigned using a three-state RIS model, such as the one described by Suter et al. [157]. That statistical weight would depend not only on the sequences of vectors on the 2nnd lattice, but also on the stereochemical sequence of the chain. The acceptance rules for use in the Metropolis criteria are then conceptually similar to those presented above in

Eqs. (4.8) and (4.9), but now based directly on the RIS model, without having to introduce coarse-grained statistical weights.

4.3.1.3
Distinguishing Features of the Two Strategies

A few comments on the trade-offs between the two approaches in Sections 4.3.1.1 and 4.3.1.2 may be useful. The approach based on the distances between 2nnd beads i–1 and i+1 can be described with arrays of smaller dimensions. Ultimately the information required for weighting of PE chains on the 2nnd lattice can be presented in a 3×3 matrix, Eq. (4.5). A much larger array, of dimensions 12×12 (Table 4.3), is required in the approach based on the orientations of pairs of vectors of length L. Two advantages accompany the use of the larger array. First, coarse-grained statistical weights [such as the a, b, and c in Eqs. (4.3), (4.4), and (4.5)] are not required in the approach based on pairs of vectors. Ambiguities in local conformations are also resolved, so that different stereochemical sequences can be weighted differently. An example of such an ambiguity is the inability to distinguish beween tg^+, tg^-, g^+t, and g^-t in the approach based on distances [the 4×4 block in Eq. (4.2)], whereas this distinction is clearly made in the approach based on the orientation of pairs of vectors of length L, Table 4.3.

The method based on the orientation of pairs of vectors of length L produces slower dynamics than the method based on next-nearest neighbor distances. Since the acceptance rates for single-bead moves differ very little in the two methods, it appears that the probability for immediate reversal of a successful move is higher in the method based on the orientation of the pairs of vectors, such that the chain "shivers" rapidly, but moves its center of mass slowly. This problem is more severe in PP melts than in PE melts. Significant diffusion of the center of mass of the chains in PP melts is achieved when reptation, as well as single-bead moves, is allowed [158, 158A].

Local collapses can be excluded from the simulation using either of the approaches described in the previous paragraph. All that is required is specification of the relationship between the 2nnd lattice and its underlying diamond lattice, so that all equilateral triangles of area $L^2/2$ can be unambiguously classified as unphysical collapses (and rigorously excluded in the simulation) or as $g^\pm g^\mp$ pairs (and permitted to the extent allowed by the RIS model appropriate for the chain under investigation.) Exclusion of collapses from the simulation slows down the translation of the center of mass and the reorientation of the end-to-end vector.

4.3.2
Intermolecular Interactions and Long-Range Intramolecular Interactions

The intermolecular interactions and long-range intramolecular interactions are incorporated via a pair-wise interaction potential. All pairs of beads from different chains are included in the *intermolecular* interactions. Of course, directly bonded pairs of beads have a configuration-independent interaction, which can be ignored in the implementation of the Metropolis criterion. The intramolecular interaction of next-nearest neighbor beads is counted in the RIS model, and is therefore ignored in the construction of the long-range interaction, in order to avoid double counting. Therefore the *long-range intramolecular* interactions apply to beads on the 2nnd lattice that are separated by three or more bonds of length L.

4.3.2.1
A Simple Approach for Θ Chains at Infinite Dilution

The potential energy function prohibits double occupancy of any site on the 2nnd lattice. In the initial formulation, which was designed for the simulation of infinitely dilute chains in a structureless medium that behaves as a Θ solvent, the remaining part of the potential energy function contains a finite repulsion for sites that are one lattice unit apart, and a finite attraction for sites that are two lattice units apart [153]. The finite interaction energies for these two types of sites were obtained by generalizing the lattice formulation of the second virial coefficient, B_2, described by Post and Zimm as [159]:

$$B_2 = \frac{1}{2}(1 - z\mu) \tag{4.14}$$

for a lattice of coordination number z, with $\mu = \exp(-u_{ij}\beta) - 1$, where u_{ij} is the potential of average force between segments i and j, and β is the reciprocal of the product of the Boltzmann constant and absolute temperature. For implementation on the 2nnd lattice, the generalization is:

$$B_2 = \frac{1}{2}(1 - z_1\mu_1 - z_2\mu_2) \tag{4.15}$$

where the coordination numbers for the first two shells, z_1 and z_2, respectively, are 12 and 42. This approach was satisfactory for the simulation of single chains at infinite dilution in a structureless Θ solvent, where the number of long-range intramolecular contacts is small [153]. The form of Eq. (4.15) requires that the attraction responsible for cohesion of a PE melt must reside entirely in the second shell. Placing all of the attraction in the second neighbor shell is not consistent with a collision diameter of 4.2–4.4 Å for $-CH_2CH_2-$ [160]. This problem with the form of Eq. (4.15) is not important in the treatment of the chain at infinite dilute in a Θ solvent, where the number of long-range intramolecular contacts is small.

4.3.2.2
A More Robust Approach for Dense Polymers

A much improved description of PE melts is obtained with a more elaborate potential energy function, based on a spherically symmetric potential energy function, u(r), acting between two isolated monomers in the gas phase [160]. Then the expression for the second virial coefficient, according to imperfect gas theory, is:

$$B_2 = -\frac{1}{2}\left\{\int [\exp(-\beta u(r)) - 1]d\mathbf{r}\right\} = -1/2\left[\int f d\mathbf{r}\right] \tag{4.16}$$

where the Mayer function replaces $\exp(-\beta u(r))$ - 1. B_2 can be rewritten in a discretized form by separating the integral into subintegrals for each 2nnd lattice cell and then regrouping them for each neighbor:

$$B_2 = -\frac{1}{2}\left[-\int_{cell} d\mathbf{r} + \sum_{1st}\int_{cell} f d\mathbf{r} + \sum_{2nd}\int_{cell} f d\mathbf{r} + \sum_{3rd}\int_{cell} f d\mathbf{r} + \ldots\right] \tag{4.17}$$

Separating out the volume element $\int_{cell} d\mathbf{r}$ of one lattice cell, V_o and introducing the cell average Mayer function, $\langle f \rangle$, defined as:

$$\langle f \rangle = \int_{cell} f d\mathbf{r} \Big/ \int_{cell} d\mathbf{r} \tag{4.18}$$

transforms Eq. (4.17) into:

$$B_2 = \frac{V_c}{2}\left[1 - \sum_{1st}\langle f \rangle_{1st} - \sum_{2nd}\langle f \rangle_{2nd} - \sum_{3rd}\langle f \rangle_{3rd} - \ldots\right] \tag{4.19}$$

$$B_2 = \frac{V_c}{2}\left[1 - z_1\bar{f}_{1st} - z_2\bar{f}_{2nd} - z_3\bar{f}_{3rd} - \ldots\right] \tag{4.20}$$

The effective interaction parameter, u_i, for the ith neighbor is:

$$\exp(-\beta u_i) - 1 \equiv \bar{f}_{ith} \tag{4.21}$$

On the 2nnd lattice, the prohibition of double occupancy of any site implies

$$u = \infty, \qquad r < 2.5\ \text{Å} \tag{4.22}$$

and u at the remaining sites can be represented by a truncated Lennard-Jones potential energy function.

$$u = 4\epsilon\left[\left(\frac{\sigma}{r}\right)^{12} - \left(\frac{\sigma}{r}\right)^{6}\right], \qquad r > 2.50\ \text{Å} \tag{4.23}$$

Table 4.5 shows the results obtained for PE at 443 K, using two slightly different combinations of ϵ and σ that are in the range expected for the interactions of two segments of -CH_2CH_2- [161]. The first shell is highly repulsive. The interaction parameter for the second neighbor shell is small, and not, apparently, negative. This interaction parameter covers the distance 2.5–5.0 Å, which includes the collision diameter, 4.2–4.4 Å. Therefore the repulsions and attractions are averaged out in the second neighbor shell. An attractive third neighbor shell is necessary for the cohesion of the system. Higher neigh-

Table 4.5 Values of u_i (kJ/mol) obtained [160] with Eq. (4.26) at 443K[a].

Shell	$\epsilon/k_B = 205\,K$, $\sigma = 4.2$ Å	$\epsilon/k_B = 185\,K$, $\sigma = 4.4$ Å
1	12.501	14.122
2	0.084	0.526
3	-0.611	-0.627
4	-0.137	-0.155
5	-0.036	-0.041

[a] Values of ϵ/k_B and σ for the Lennard-Jones potential function, Eq. (4.23), are reported to be 205 K and 4.2 Å for ethylene, and 230 K and 4.4 Å for ethane [161].

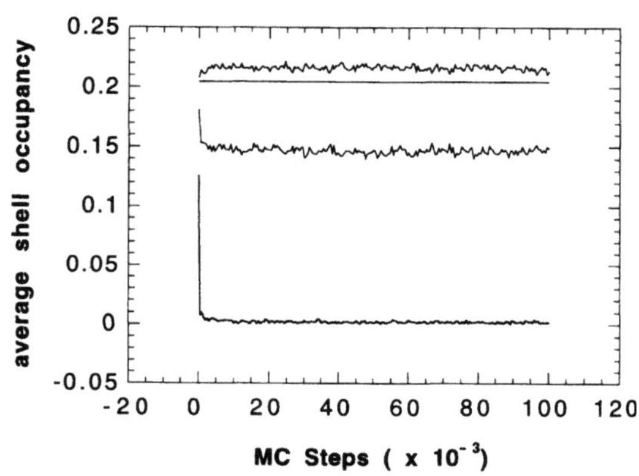

Fig. 4.4. Equilibration of the occupancies of the first three shells in a PE melt

bors are also attractive, but less so than the third neighbor. The simulations described here for PE melts were performed using only u_1, u_2, and u_3 in evaluating the probabilities for acceptance of a single bead move. In the subsequent analysis of individual replicas for PE, as in the computation of a cohesive energy, the tail of the potential energy function, represented by u_4 and u_5, may be incorporated.

Constant volume simulations performed for melts of $C_{100}H_{202}$ with the first set of parameters in Table 4.5, and at various densities (imposed upon the simulation by the selection of the number of chains and the dimensions of the periodic box), show rapid equilibration of the occupancies of the various shells [160]. At 443 K and a density of 0.853 g/cm^3, the occupancy of the first three neighbor shells reaches a plateau in a few thousand MC steps, as depicted in Fig 4.4. The horizontal line in this figure is drawn at 0.205, which is the fraction of the sites on the 2nnd lattice that must be occupied in order to produce a density of 0.853 g/cm^3. The first neighbor shell is rarely occupied, the occupancy of the second nearest neighbor shell is about 3/4 of the average occupancy, and the occupancy of the third shell is slightly higher (by a few percent) than the average occupancy for the lattice as a whole. Occupancies of the fourth and higher neighbor shells are closer to 0.205, which is the average occupancy. The PE melt selects a density that is in a reasonable range, when given the option [162].

The cohesive energy per monomer, evaluated at a density of 0.853 g/cm^3 using the first potential energy function described in Table 4.5, is 7.0 kJ/mol, using u_1–u_5 [160]. The experimental cohesive energy at ambient temperature is in the range 8–9 kJ/mol [163]. Refinements of the potential energy function, with the objective of matching more closely the experimental physical densities and cohesive energies with the coarse-grained system on the 2nnd lattice, can be achieved by two routes. The first route simply employs minor adjustments (larger ϵ and/or larger σ) in the Lennard-Jones potential from which the effective interaction parameters for the various shells are derived. The second route provides a more elaborate description of the interactions on the 2nnd lattice, while retaining a specified ϵ and ω. This second route increases the number of shells included in the calculation, thereby including more of the long attractive tail of the Lennard-Jones potential. It also separates distinctly different vectors for the various shells that are lumped together in Eq. (4.17). For example, the second neighbor shell has six types of vectors. Five types (1,–1,1; 1,1,0; 2,–1,–1; 2,0,0; 2,–2,0) are represented by six spherically equivalent vectors, and the sixth type (2,–1,0) by twelve spherically equivalent vectors, for a total of 42. The range of distances covered by these six types is different, and they therefore yield different values for the average Mayer function f via Eq. (4.18). Using $\epsilon/k_B = 205$ K, $\sigma = 0.42$ nm, at 443 K, the u for these six types of vectors in the second neighbor shell ranges from –0.5969 to 0.3754 kJ/mol, and an average over all six types of vectors is used for the evaluation of u_2 when the interaction potential is based on neighhbor shells, as in Table 4.5. A more accurate result for properties such as the preferred density and cohesive energy might be obtained by expressing the interaction potential using types of vectors, rather than types of shells.

The procedure in Eqs. (4.16)–(4.23) has been employed to describe the long-range interactions for the simulation of PP melts [158, 158A]. Bulk density is achieved with occupancy of 12% of the sites on the 2nnd lattice, with each occupied site representing a C_3H_6 unit. As expected, the long-range potential for PP is shifted to larger distances than is the case for PE (Table 4.5). The first shell is strongly repulsive for both polymers. The second shell is definitely repulsive (u_2 = 5.1 kJ/mol) for PP, and nearly equal attractions are obtained in the third and fourth shells (u_3 = –0.81 kJ/mol, u_4 = –0.77 kJ/mol). Subsequent shells are less strongly attractive (u_5 = –0.23 kJ/mol ... u_9 = –0.007 kJ/mol). The first four shells were employed during the simulation of PP, in making the decision for acceptance of a move. However, the first nine shells were employed in the computation of the cohesive energies, which were 14.57 and 14.53 kJ/mol for melts of isotactic and syndiotactic PP, respectively. Slightly smaller values (13 and 13.77 kJ/mol) were tabulated by Bicerano [164]. For isotactic and syndiotactic PP melts, the cohesive energy density evaluated considering only the first four shells of the long-range interactions is only about 2/3 of the results obtained using the first nine shells [158].

The strength of the motivation for developing extremely accurate formulas for the potential energy function for the intermolecular and long-range intramolecular interactions on the 2nnd lattice is coupled with the role that Step 3 will play in the overall simulation. The greater interest may be in the properties of the replicas obtained after restoration of the missing atoms, which will be described below. Then it may be sufficient to have a reasonable approximation to the "best" potential in Step 2, with the final refinement in the description of the system to be obtained after the completion of the "reverse mapping" back to the fully atomistic description of the system in continuous space, in Step 3.

4.4 The Third Step – Reverse Mapping

Reverse mapping restores all of the missing atoms, and takes the system from the discrete space of the 2nnd lattice to a continuum. The starting point for reverse mapping depends on choices made in the performance of the second step.

4.4.1 *Resolution of Collapsed Beads*

For a simulation of PP, the relationship between the 2nnd lattice and its underlying diamond lattice must be established at the beginning of the simulation, in order to preserve the stereochemical sequence and its influence on the conformations of the chains. The half of the equilateral triangles of area $L^2/2$ that produce local collapsed beads is therefore known at the start, and the simulation can be performed in a manner which avoids the formation of these unphysical structures [158].

The simulation of PE, on the other hand, can be carried out in the second step without specification of which of the two underlying diamond lattices will ultimately be selected for the reverse mapping, as discussed in Sect. 4.3.1.1. If this was the case in Step 2, one of the two potential underlying diamond lattices is selected at random at the beginning of Step 3. Then a very short trajectory is run, with assignment of a high energy to the collapses (which are now unambiguously identified) in order to remove them from the system. For simulations using nine chains of $C_{100}H_{202}$ at 0.75 g/cm^3 and 473 K, all collapses were removed in 10–140 MC steps, depending on the specific snapshot [144]. In simulations of 25 chains of $C_{316}H_{634}$ at 473 K and 0.76 g/cm^3, on average 50 such collapses (an average of two per chain) are observed in each snapshot [144], which is comparable with the number of $g^{\pm}g^{\mp}$ pairs expected from the RIS model at this temperature [143].

In principle, another type of collapse, of intermolecular rather than intramolecular origin, might occur whenever four beads (two from one chain, and two from another chain) describe a regular tetrahedron, with side L, on the

2nnd lattice. Then reverse mapping may select the site at the center of mass of the tetrahedron for the carbon atoms that are introduced between the two 2nnd beads for both pairs of chains. Although this mechanism of collapse is conceivable, it has never been observed in any of our simulations of PE. The explanation for its absence can be found in the intermolecular energy of the four beads on the 2nnd lattice when they describe the required regular tetrahedron. There will be four intermolecular interactions with energies u_1, where u_1 is in excess of 12 kJ/mol (see Table 4.5). The energy of $\sim$ 50 kJ/mol for formation of the regular tetrahedron with side L apparently accounts for its absence in our simulations of PE.

Avoidance of the intramolecular collapses in PE can be achieved by selection of a relationship between the 2nnd lattice and its underlying diamond lattice at the outset of the simulation, and then prohibition of occupation by three beads of that half of the equilateral triangles with area $L^2/2$ that form the local collapse. Implementation of this restriction did not decrease the acceptance rate significantly (the decrease is less than 1%) in our simulations of amorphous PE at bulk density. The acceptance rate is scarcely affected because the moves that would have created the intramolecular collapse were already of low probability, due to their weighting by ω, which is much less than one. There was, however, a reduction in the mean square displacement of the center of mass per MC step, as discussed in Sect. 4.3.1.3.

4.4.2
Restoration of the Missing Atoms

After removal of the collapsed beads (if any were present), all carbon and hydrogen atoms are restored, using the underlying diamond lattice, standard bond lengths and tetrahedral bond angles. The coordinates and connectivity of all carbon and hydrogen atoms are transferred to another software package for a rapid minimization in the potential energy of the system. The RIS model and the potential energy functions that were derived for the 2nnd lattice are no longer required. Instead the simulation switches to another force field that is designed for an accurate description of the interactions in the fully atomistic system, and permits the coordinates of the atoms to move into continuous space. Commercial software (Discover 3.0 with the CVFF force field, version 2.3, Molecular Simulations, Inc., no cross terms and no Morse potentials) has been successfully used for this purpose [144]. The initial stage in the energy minimization uses nonbonded energies scaled by 10^{-5}, to relieve initial close contacts between hydrogen atoms. Successive stages alter the scaling by factors of ten, until the full nonbonded potential energy functions are used, and the gradient is less than 0.1 kcal/(mol Å). An illustrative snapshot of the result is depicted in Fig. 4.5.a and 4.5.b. Only a rapid energy minimization (but no computationally intensive MD) is performed with the fully atomistic representation of the system.

a

b

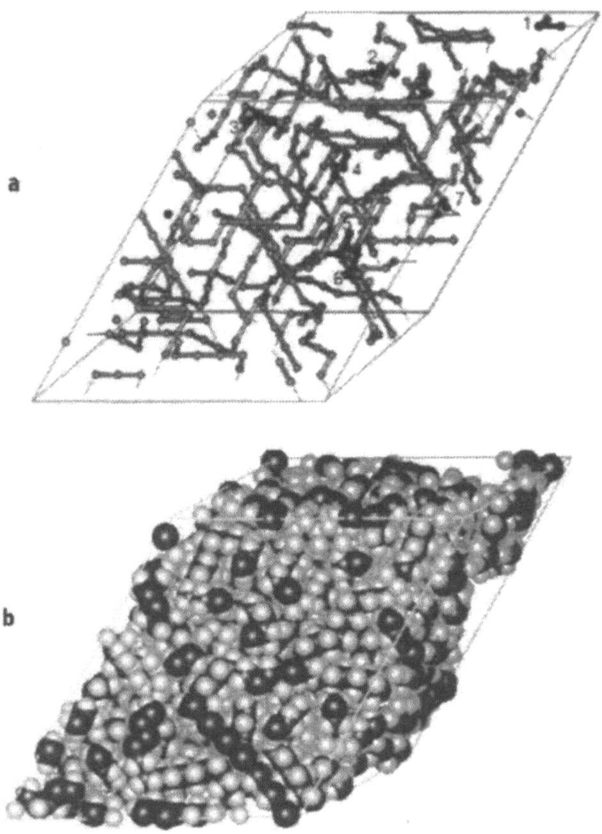

Fig. 4.5. Periodic cell for nine chains of $C_{100}H_{202}$ at 0.75 g/cm^3 and 443 K. **a** On the 2nnd lattice, with numbers denoting the locations of collapsed beads and **b** After completion of the reverse mapping in the third step. Adapted from [144]

4.4.3
Some Properties of the Regenerated Fully Atomistic Replicas

The success of reverse mapping can be assessed by examination of local conformations, intermolecular pair distribution functions, and the overall energetics of the system. After the initial introduction of all missing atoms, but before the energy minimization, the distribution of the torsion angle at the internal CH_2–CH_2 bonds is described by delta functions located precisely at $\phi = \pm 60$ and 180°. An illustrative result for this distribution after the completion of the energy minimzation is depicted in Fig. 4.6 [144]. Before taking the system into continous space, the delta functions in this specific snapshot had heights of 0.624 at 180°, and 0.179 and 0.197 at $\pm 60°$. The distribution in continuous space retains the t and $g^{\pm}$ rotational isomeric states, but broadens the distributions of ϕ within each state. The maximum probability for the t

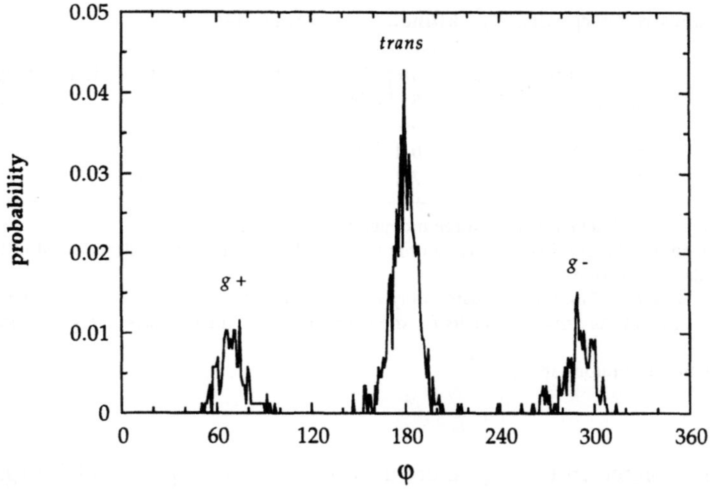

Fig. 4.6. Distribution of torsion angles at the CH_2–CH_2 bonds for the snapshop depicted as **b** in Fig. 4.5. Adapted from [144]

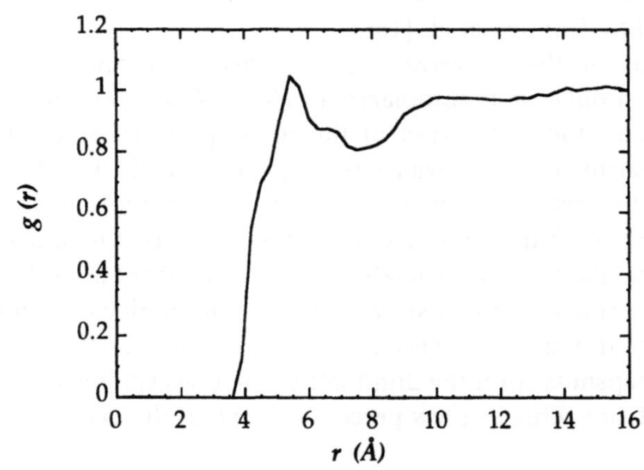

Fig. 4.7. Intermolecular carbon-carbon pair distribution functions for $C_{100}H_{202}$ after completion of the third step. Adapted from [144]

state remains at 180°, but the maxima in the distribution for the g states moves from precisely ±60° in the discrete space of the lattice to $\pm(60°+\Delta\phi)$ in continuous space. The value of $\Delta\phi$ appears to be in the range 0–10°, consistent with the energetic analysis performed by Abe et al. [143].

An intermolecular pair distribution function evaluated at the end of Step 2 would consist of delta functions at those distances allowed on the 2nnd lattice. After completion of reverse mapping, which moves the system from the discrete space of the lattice to a continuum, the carbon–carbon intermolecular pair distribution function becomes continuous, as depicted in Fig. 4.7 [144].

Table 4.6. Cohesive energy density,[a] kJ/mol, for $C_{100}H_{202}$ [144]

Density	Step 2, 2nnd Lattice, 3 shells[b]	Step 2, 2nnd lattice, 5 shells[c]	Completion of Step 3[d]
0.75 g/cm^3	4.0 ± 0.1	6.4 ± 0.1	8.5 ± 0.2
0.86 g/cm^3	4.6 ± 0.1	7.4 ± 0.0	9.9 ± 0.1

[a] Average and standard deviations for three independent replicas.
[b] Coarse-grained, on the 2nnd lattice, using only the first three shells for the description of the intermolecular interactions.
[c] Coarse-grained, on the 2nnd lattice, using only the first three shells during the MC simulation in Step 2, but employing the first five shells for the computation of the cohesive energy for each snapshot.
[d] Fully atomistic, continuous space.

The results depicted in the figure are averages for 10 snapshots at 0.75 g/cm^3. Similar positions and amplitudes for the first maximum and first minimum were obtained in MD simulation for a $C_{44}H_{90}$ melt at 400 K and 0.76 g/cm^3 [165]. The kink near 4 Å does not appear in this MD simulation, but a similar kink does appear in the site–site intermolecular radial distribution function for PE reported by Honnell et al. [166].

The behavior of the cohesive energy density for amorphous $C_{100}H_{202}$ throughout the simulation is summarized in Table 4.6. Somewhat larger values are obtained after the restoration of the missing atoms and completion of Step 3. The experimental cohesive energy reported for PE at ambient temperature, for which the last row of entries in Table 4.6 is appropriate, is in the range 8–9 kJ/mol [163]. With the parameters that were used on the lattice, and using only the first five shells in the evaluation of the attractive tail of the long-range interactions, the coarse-grained system on the 2nnd lattice does not achieve a cohesive energy this high. However, with the aid of the reverse mapping in Step 3, these snapshots from the 2nnd lattice yield acceptable cohesive energy densities, which are actually a few percent larger that the expected values.

4.5 Dynamics

We now turn to a consideration of the dynamics observed for the coarse-grained system during the execution of Step 2. This step might be employed for two distinctly different types of problems. Perhaps the ultimate interest is in the simulation of static properties of a completely equilibrated amorphous system at bulk density. Then the dynamic run on the 2nnd lattice would have the objective of removing any stresses that were in the system as initially constructed. The analysis would not consider the dynamics as computed in Step 2, but would instead focus attention only on the equilibrated coarse-grained systems on the 2nnd lattice, or the fully atomistic systems obtained after completion of the reverse mapping in Step 3.

Alternatively, it might be the dynamics itself that is of interest, as would be the case in the study of properties that depend on the translational diffusion coefficient or the relaxation time for the reorientation of the end-to-end vector. This alternative requires a higher standard for the behavior of the simulation during Step 2 than does an investigation that only examines static properties of completely relaxed systems. Figures 4.8 and 4.9 depict the mean square displacement of the center of mass and the decorrelation of the end-to-end vector, $\langle \mathbf{r}(t) \cdot \mathbf{r}(0) \rangle / \langle \mathbf{r}^2 \rangle$, respectively, for melts of $C_{316}H_{634}$ at 473 K [167]. The number of chains, dimensions of the periodic box, density, and occupancy of the sites on the 2nnd lattice employed in these two simulations, as well as another simulation using $C_{44}H_{90}$, are listed in Table 4.7. The $\langle R_G^2 \rangle_0$ for $C_{316}H_{634}$ is denoted by the horizontal arrow in Fig. 4.8. Its location shows that the system has been followed in Step 2 for times well above the time required for

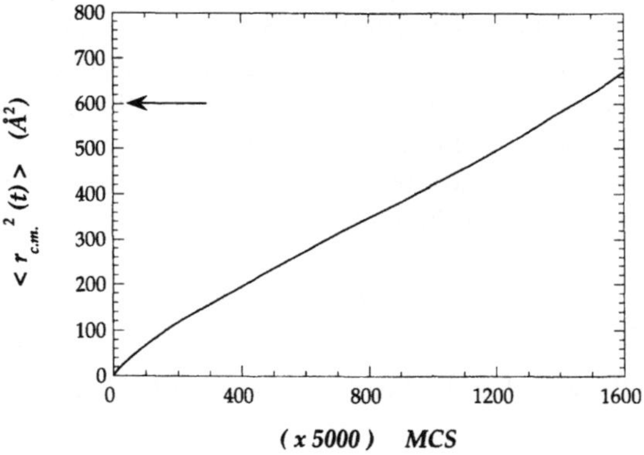

Fig. 4.8. Mean square displacement of the center of mass during the simulation of $C_{316}H_{634}$ [167]. The *horizontal arrow* denotes the mean square unperturbed radius of gyration

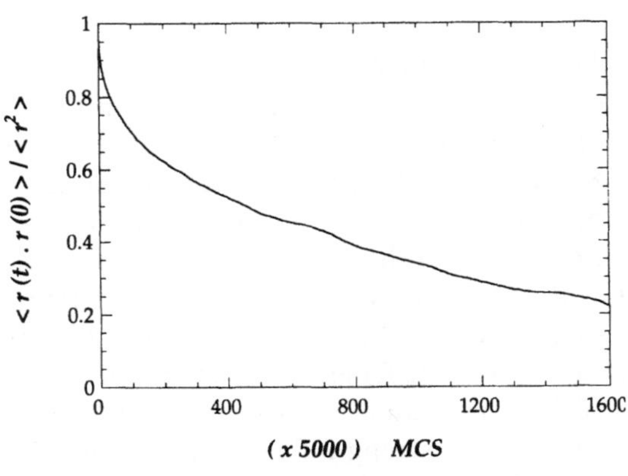

Fig. 4.9. Decorrelation of the end-to-end vector during the simulation of $C_{316}H_{634}$ in Fig. 4.8 [167]

Table 4.7. Periodic boxes employed for simulations of two n-alkane melts [167]

n-Alkane	Number of chains	Periodic box (Å)[a]	Density (g/cm^3)	Occupancy
$C_{44}H_{90}$	19	32.5 × 35.0 × 35.0	0.692	0.164
$C_{316}H_{634}$	25	70.0 × 70.0 × 70.0	0.758	0.180

[a] Measured along the axes of the unit cell on the 2nnd lattice, see Fig. 4.2.B.

diffusion of the center of mass over a distance equal to the root-mean-square radius of gyration. This statement is also true for the simulation of $C_{44}H_{90}$ (not depicted in Fig. 4.8), where Step 2 was an order of magnitude longer than the time required for diffusion of the center of mass over a distance equal to its $\langle R_G^2\rangle_0^{1/2}$.

MD simulations of melts of $C_{44}H_{90}$, based on classic techniques in continuous space, have been reported recently using united atom [146] and fully atomistic [145] representations of the chain. Time in the conventional MD simulations is expressed in seconds, whereas time in the simulation of the coarse-grained chains on the 2nnd lattice is expressed in MC steps. Nevertheless, a few comparisons are possible via the longest relaxation time, τ_r, deduced from the decorrelation of the end-to-end vector:

$$\frac{\langle \mathbf{r}(t)\cdot\mathbf{r}(0)\rangle}{\langle r^2\rangle} = \sum_{p=1,3,\ldots} \frac{8}{p^2\pi^2}\exp(-tp^2/\tau_r) \tag{4.24}$$

and the friction coefficients for the beads, ζ, deduced from τ_r:

$$\zeta(\tau_r) = \frac{3\pi^2\tau_r}{\beta N\langle r^2\rangle} \tag{4.25}$$

and from the translational diffusion coefficient for the center of mass:

$$\zeta(D_N) = \frac{1}{\beta N D_N} \tag{4.26}$$

The differences in the units for time in the classic MD simulation and in the simulation of the coarse-grained chain during Step 2 on the 2nnd lattice inhibit a direct comparison of the values of τ_r and D_N, which are nevertheless reported in Table 4.8. This difficulty can be eased by manipulation of the data in a manner such that the quantities examined no longer have units that include time, as in the formation of the product $\tau_r D_N$ and the ratio $\zeta(\tau_r)/\zeta(D_N)$. These two quantities are also reported in Table 4.8. The ratio $\zeta(\tau_r)/\zeta(D_N$ is particularly interesting, because intuition suggests that this ratio should be very close to 1. The last column in Table 4.8 shows that this ratio is larger than 1 in the conventional MD simulation, perhaps because $C_{44}H_{90}$ is sufficiently short so that end effects have a strong influence, as suggested by the authors [145,146]. This ratio is at least as close to 1 for the MC simulation of the coarse-grained chains on the 2nnd lattice as it is for the simulations by more

Table 4.8. Comparison of atomistic and 2nnd simulations of $C_{44}H_{90}$ melts

Simulation	τ_r	D_N	$\tau_r D_N$	$\zeta(\tau_r)/\zeta(D_N)$
Fully atomistic[a]	1614 ps	1.9×10^{-2} Å^2/ps	30.7	1.23
United atom[b]	938 ps	3.5×10^{-2} Å^2/ps	32.8	1.29
Set I[c,d]	8.5×10^4 MCS	2.0×10^{-4} Å^2/MCS	17.1	1.00
Set II[c,e]	1.2×10^5 MCS	1.5×10^{-4} Å^2/MCS	17.9	1.00
Set II[c,f]	1.7×10^6 MCS	1.7×10^{-5} Å^2/MCS	28.6	1.25

[a] Classic MD in continuous space, as reported by Smith et al. [145]
[b] Classic MD in continuous space, as reported by Paul et al. [146]
[c] Coarse-grained chains on the 2nnd lattice, using three (u_1, u_2, u_3) shells for the intermolecular potential energy function employed in the Metropolis criteria during the MC simulation [167]
[d] u_i from the first set in Table 4.5
[e] u_i from the second set in Table 4.5
[f] u_i from the second set in Table 4.5, and with the RIS parameters changed so that the mean square dimensions match more closely those obtained by Smith et al. [145] and Paul et al. [146]

conventional MD techniques in continuous space. The exact value of the ratio is affected by the choice of united atom vs. fully atomistic representations in the conventional MD simulations, and by parameterization of the RIS model used in the MC simulations on the 2nnd lattice.

Comparison of MD and MC simulations for longer chains has not yet been performed. MD simulations of melts of $C_{100}H_{202}$ are available [168]. Figures 4.8 and 4.9 show that the necessary simulations for melts of longer chains, such as $C_{100}H_{202}$ and $C_{316}H_{634}$, can be performed via an MC simulation on the 2nnd lattice.

Calibration of the MC time step in the simulation on the 2nnd lattice can be achieved by comparison of τ_r or D_N with the results from a conventional MD simulation (as in the second and third columns of Table 4.8), or via comparison with a translational diffusion coefficient obtained from experiment with a real system [169].

The computational advantage of the MC simulation on the 2nnd lattice, compared with a conventional MD simulation, depends on the conversion between MC steps and time measured in seconds, and on the ratio of CPU-time required for execution of a single MC step and execution of a single MD step. Comparison with a conventional MD simulation, as implemented in a popular commercial software package, has been achieved by performing a pair of simulations on the same SGI workstation. The simulations employ the same small system (nine chains of $C_{100}H_{202}$ at a density of 0.86 g/cm^3). The overall result was that the MC simulation of the system on the 2nnd lattice was faster, by a factor of 360 ± 80, than the conventional MD simulation.

For *n*-alkane melts on the 2nnd lattice, the time for computation of a single MC step scales approximately as $N^{1.1}$, where N denotes the number of beads. This exponent was evaluated for simulations with 1500, 2844, and 3950 beads. The exponent for N seems to decrease slightly as N increases; it may be closer to 1 for larger systems.

4.6
Other Types of Systems

The systems described above have employed monodisperse PE or PP melts at bulk density. Several variations on this theme can be studied by the same technique. When the PP chains have different stereochemical compositions, the simulation provides information about their miscibility in the melt [158A]. The chains may be polydisperse, thereby providing access to simulations of the diffusion of probe chains through a matrix composed of chains of a different size [170]. Isolated cyclic PEs have been simulated successfully [153]. The cyclic PEs were easily examined at bulk density, either as pure species or in combination with linear chains. The cyclic PEs were also prepared and studied as intertwined "Olympic" rings. PE with branches larger than methyl groups have not yet been studied on the 2nnd lattice, but the availability of an accurate RIS description for PE star-branched polymers [171] strongly suggests that the branched species could easily be mapped onto the 2nnd lattice. The ability to treat branched and cyclic PE carries with it the implication that the method can easily be extended to crosslinked networks.

The method described by Misra et al. [172] for transforming a fully atomistic model of an amorphous polymer at bulk density into a model for the surface presented by that polymer to a vacuum can easily be applied to the simulations of the coarse-grained polymer on the 2nnd lattice [162], as depicted in Fig. 4.10. The simulation produces excellent values for the density profile and surface energy of PE melts, using the method of Misra et al. [172], even when the thin films are in the coarse-grained representation on the 2nnd lattice. Thin films with thickness in excess of 100 Å have been prepared and equilibrated. Surface energies for the PE melts are independent of thickness for films thicker than 40 Å [162]. The dynamics at the segmental level and at the

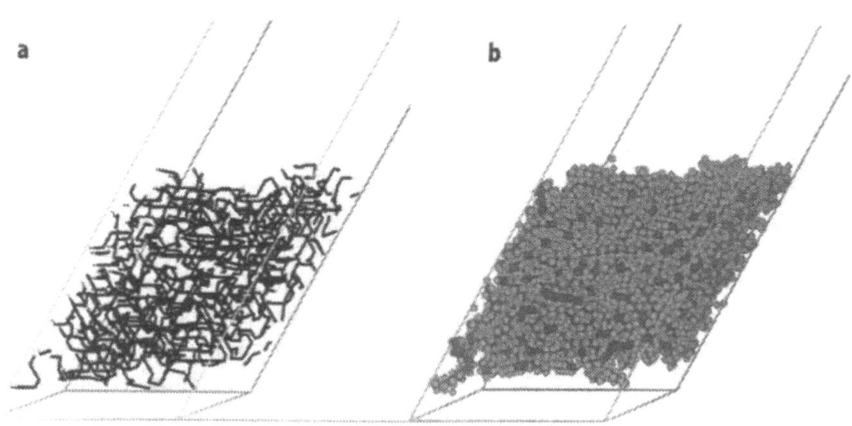

Fig. 4.10. A thin film of a melt of $C_{100}H_{202}$, with both surfaces presented to a vacuum, as represented **a** on the 2nnd lattice and **b** after reverse mapping. Adapted from [162]

level of entire chains have been evaluated for these films, as a function of the distance of the chain from the nearest surface [167]. Dense brushes formed by the attachment of the ends of specific *n*-alkanes to a surface, should also be accessible, both with regard to their equilibrium properties and their dynamics.

The computational efficiency of the dynamics in Step 2 on the 2nnd lattice permits the study of the cohesion (and complete mixing) of two thin films [173]. Time scales for equilibration of the density profile, redistribution of chain ends, and complete intermixing of the chains span several orders of magnitude of time, expressed in MC steps, and are all accessible via the 2nnd simulation [173].

Isolated unperturbed polyoxyethylene chains have been simulated on the 2nnd lattice [154]. The literature contains RIS models for a large number of polyethers [124], and it is likely that most of these chains could be mapped onto the 2nnd lattice with little difficulty. It is also likely that the work on PP [156,158] can be extended to other vinyl polymers, such as poly(vinyl chloride). This capability should permit the construction and complete equilibration of amorphous poly(vinyl chloride) cells larger than those described to date. They may be large enough to address issues arising from the weak crystallization reported for these systems [174].

5 Mapping Atomistically Detailed Models of Flexible Polymer Chains in Melts to Coarse-Grained Lattice Descriptions: Monte Carlo Simulation of the Bond Fluctuation Model

5.1 Introduction: General Concepts

As described in Sect. 1, the relevant length scales and time scales are a serious problem for any simulation of polymer melts [12,16–20] and, as discussed, a polymer coil has structures on different distance scales (Fig. 1.2) [17] and relaxations on different time scales. A brute force approach, consisting of a simulation of fully atomistic models of a sufficiently large system over time scales for which thermal equilibration could be reached at practically relevant temperatures, is totally impossible. Useful progress requires a different approach.

Any approach different from this brute force approach must make compromises, as far as the complete realistic modelling of polymeric materials with all their details is concerned. Different groups tend to make rather different compromises, depending on what features of the problem they consider particularly important. Here we discuss only one approach proposed [28,30, 32,175,176] by the condensed matter theory group at the University of Mainz. This approach follows a rather radical concept, since all fast vibrational motions are completely eliminated, and in addition a description of the local

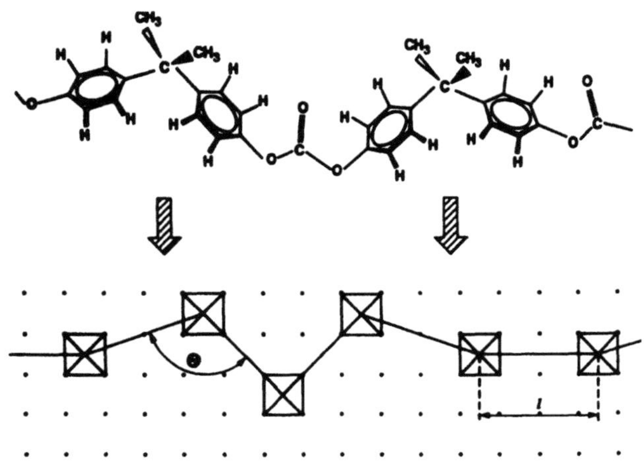

Fig. 5.1. A chemically realistic description of a polymer chain (bisphenol-A-polycarbonate [BPA-PC] in the present example) is mapped approximately onto the bond fluctuation model, by using suitable potentials for the length ℓ of the effective bonds and the angles θ between them. In this example (3 : 1 mapping) one chemical repeat unit of BPA-PC containing n = 12 covalent bonds along the backbone of the chain is translated into three effective bonds. From [28]

chemical and geometrical structure on small scales (less than about a few Å) is disregarded. The idea is to attempt a "mapping procedure" of the chemically realistic model of a polymer to equivalent chains of a lattice model (Fig. 5.1) [28]. Of course, such a mapping is never fully exact, and there is also a freedom in the choice of the precise mapping ratio (how many covalent bonds of the chemically realistic chain are integrated into one effective bond of the coarse-grained chain). The most suitable choice of this ratio has to be studied separately for each polymer type. The chemical structure and geometry of the actual repeat units, as well as their energetics (described by potentials for bond lengths, bond angles, torsional angles), enter the problem in a rather indirect way, by choosing effective potentials $U(\ell)$ for the effective bond length, $V(\vartheta)$ for the angle ϑ between them, etc. The stochastic transitions between the minima of the torsional potential (Fig. 1.3) are also modelled by stochastic hopping events of the effective monomers on the lattice (Fig. 5.1). The transition probabilities for these random hopping moves have to take the effective potentials mentioned above into account, as always in Metropolis-type Monte Carlo (MC) algorithms [12,19,20]. However it is now important to realize that the attempt frequency with which motions are tried in the MC algorithm is defined only up to a time scale factor – the mapping of this (artificial) MC time scale (traditionally measured in units such as 1 Monte Carlo step per monomer = 1 MCS) to the physical time units (seconds) is a subtle problem [28,30,32,175,176].

This mapping by which real polymers and their chemical details are forced from the continuum onto the lattice, seems reminiscent of Procruste's bed of

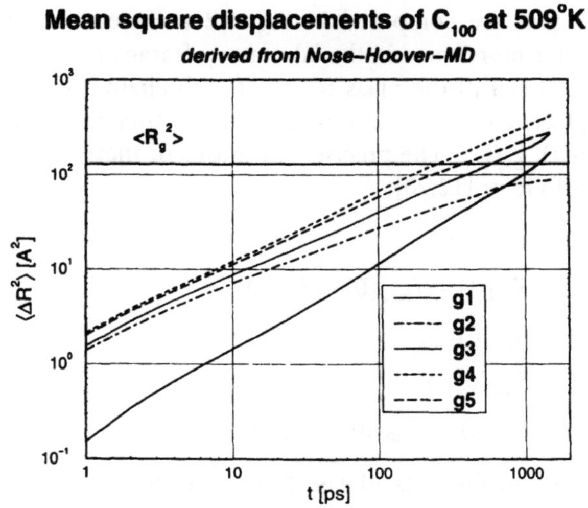

a

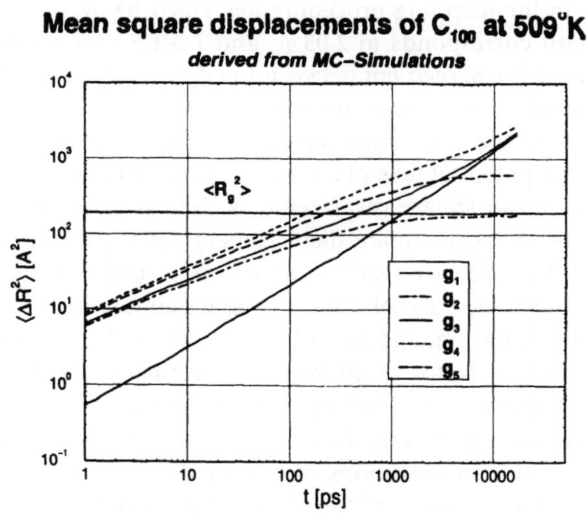

b

Fig. 5.2. **a** Mean square displacements $g_1(t)$, $g_2(t)$, $g_3(t)$, $g_4(t)$, $g_5(t)$ [Eqs. (5.1)–(5.5)] of PE (for $N_p = 100$) plotted vs. time (in picoseconds) at a temperature of 509 K. Data are taken from the molecular dynamics simulations in [168]. **b** Mean square displacements as above, but obtained from the MC simulation of the coarse-grained lattice model. From [32]

Ancient Greek mythology (legs of travellers that were too short or too long were forcibly stretched or cut until they fitted) – thus one may have some doubts that this approach has any reasonable chance at all.

Such doubts are removed by a comparison of results for polyethylene (PE), (Fig. 5.2) obtained at such high temperatures (namely T = 509 K), that it is already possible to carry out molecular dynamics simulations for melts of (sufficiently short but chemically realistic!) chains that reach thermal equilibrium [165,168].

The quantities that are compared in Fig. 5.2 are mean square displacements, $g_1(t)$, of inner monomers in the laboratory frame and analogous quantities, $g_2(t)$, in the center of the mass frame of each chain, the center of mass mean square displacement, $g_3(t)$, and mean square displacement of monomers at chain ends, $\{g_4(t), g_5(t)\}$. The precise definitions of these mean square displacements are as follows [12,20]:

$$g_1(t) = \langle [\vec{r}_{N/2}(t) - \vec{r}_{N/2}(0)]^2 \rangle, \tag{5.1}$$

$$g_2(t) = \langle [\vec{r}_{N/2}(t) - \vec{r}_{CM}(t) - \vec{r}_{N/2}(0) + \vec{r}_{CM}(0)]^2 \rangle, \tag{5.2}$$

$$g_3(t) = \langle [\vec{r}_{CM}(t) - \vec{r}_{CM}(0)]^2 \rangle, \tag{5.3}$$

$$g_4(t) = \langle [\vec{r}_{end}(t) - \vec{r}_{end}(0)]^2 \rangle, \tag{5.4}$$

$$g_5(t) = \langle [\vec{r}_{end}(t) - \vec{r}_{CM}(t) - \vec{r}_{end}(0) + \vec{r}_{CM}(0)]^2 \rangle. \tag{5.5}$$

There was no fitting to the data of Fig. 5.2(a) involved in the scales of distances and times for the lattice model; instead, the scale factors were simply used as found from the mapping procedure described below, namely (for T = 509 K) 1 lattice unit corresponds to 2.03 Å, and 1 MCS corresponds to 20 fs. Therefore the reasonable agreement between the data of Fig. 5.2a and Fig. 5.2b is a significant result, the comparison did not involve any adjustable parameters whatsoever. At the high temperature of 509 K, the jumps over the barriers of the torsional potential take place on a time scale of about 7 ps, and the log-log plot of the mean square displacements against time shows that after roughly a nanosecond mean square displacements of the order of the mean square gyration radius $\langle R_g^2 \rangle$ are reached, a necessary condition for good thermal equilibrium. However, the molecular dynamics simulation took about 100 hours on a RISC work station, and, in order to obtain high precision data, one needs a run that lasts a factor of 10 longer. In contrast, the MC results do extend a decade longer in time, but needed only 10 hours computing time! Part of this gain factor is due to the fact that lattice algorithms perform faster since less actual calculation operations are required; but more important even is the fact that n = 5 C–C bonds of the chemically realistic chain correspond to one effective bond of the bond fluctuation model; on the lattice we hence simulate PE with $N_p = 100$ by a chain with chain length $N = N_p/n = N_p/5 = 20$. The true computational advantage of the MC model, however, only becomes apparent at lower temperatures, where the barrier of the torsional potential is more strongly felt — and in the MC approach this slowing down can be incorporated to a significant extent in the conversion factor between 1 MCS and the physical time scale, as we shall see below.

Note that the accuracy with which the MD calculation can estinate the gyration radius is only about 10%, and thus it is not clear whether the slight disagreement between the MC and MD results for $\langle R_g^2 \rangle$ in Fig. 5.2 is significant. We empharize a comparison of dynamic properties here because the usefulness of MC to estinate any dynamic properties of polymers is doubled often in the literature. Comparisons her static properties on smaller length scales –

which are more early obtained accurately by both methods – are shown in Fig. 5.8a,b below.

We also emphasize that the MD model does include the vibrational motions of bond, and torsional angles (in the minima of the respective potentials) but, somehow, these small scale fast motions are rapidly damped out in the melt, and do not affect the motion on the nanometer scale (and for corresponding times) significantly.

5.2 The Bond Fluctuation Model of Dense Polymer Melts: A Brief Review of Its Properties

At this point, we recall the general features of the athermal bond fluctuation model [37–42] where no potentials for bond lengths $U(\ell)$, bond angles $V(\theta)$ etc. have yet been built in. We use a simple cubic lattice, and the effective monomers are not just described by single lattice points as in the standard self-avoiding walk model [16] of polymer chains, but rather each effective monomer blocks all eight corners of an elementary cube of the lattice from further occupation. The effective bonds connecting the effective monomers can have the lengths 2, $\sqrt{5}$, $\sqrt{6}$, 3, $\sqrt{10}$ lattice spacings. (The length $\sqrt{8}$ is forbidden, since then the random hopping algorithm can never lead to an intersection of chains, which otherwise would be possible: thus excluded volume interactions automatically ensure entanglement constraints.) This leads to a total of 108 possible next-neighbor pairs of which about 12 can be occupied simultaneously. Thus this model shows the feature of existing on a high coordination lattice as discussed in Sect. 4. In the course of the dynamic MC simulation, a monomer is selected at random for a jump trial in a randomly selected lattice direction (by one lattice unit). This jump only takes place if the two bonds connecting the jumping monomers to its neighbors still belong to the set of allowed values after the jump, and if no violation of excluded volume interactions occurs. [In the case of the thermal model, changes in potentials (for bond lengths, bond angles, etc.) may also lead to an energy change $\Delta\mathcal{H}$ in such a trial. If $\Delta\mathcal{H} > 0$, a transition probability $P = \exp(-\Delta\mathcal{H}/k_BT)$ needs to be calculated, and a random number ζ_r uniformly distributed between zero and unity is drawn. The move is then only carried out if P exceeds ζ_r.] In order to simulate melts, a volume fraction $\Phi = 0.5$ of occupied sites (or somewhat larger) is needed, since then the excluded volume interaction is effectively screened on the scale of the length of an effective bond, and chains have nearly Gaussian-type configurations [40].

Although this athermal bond fluctuation model is clearly not yet a model for any specific polymeric material, it is nevertheless a useful starting point from which a more detailed chemical description can be built. This fact already becomes apparent, when we study suitably rescaled quantities, such that, on this level, a comparison with experiment is already possible. As an example, we can consider the crossover of the self-diffusion constant from Rouse-like behavior for short chains to entangled behavior for longer chains.

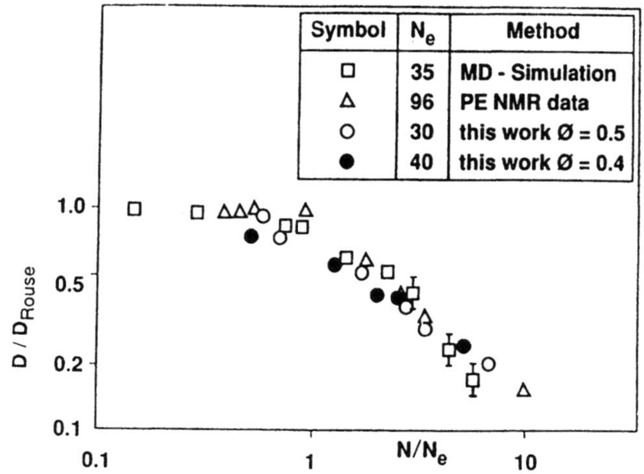

Fig. 5.3. Log-log plot of the self-diffusion constant D of polymer melts vs. chain length N. D is normalized by the diffusion constant of the Rouse limit, D_{Rouse}, which is reached for short chain lengths. N is normalized by N_e, which is estimated from the kink in the log-log plot of the mean-square displacement of inner monomers vs. time [$g_1(t)$ vs. t]. Molecular dynamics results [177] and experimental data on PE [178] are compared with the MC results [40] for the athermal bond fluctuation model. From [40]

Plotting the self-diffusion constant, D_N, in units of its limiting behavior (D_{Rouse}) for short chains and rescaling the chain length N in terms of the entanglement chain length N_e, one obtains Fig. 5.3 [40]. Here molecular dynamics results [177] and experimental data on PE [178] are included and, within the accuracy of the simulation data, they all fall on a universal curve, i.e. material-dependent prefactors (in the experiment) or model-dependent prefactors (in the simulation) are absorbed into the normalization of units of the coordinate scales. Accepting the suggestion [28,30,32,37–42,175,176] that the experimental degree of polymerization (N_p) should be related to the number N of effective monomers as $N_p = nN$, with n = 3–5, it is also very gratifying that the experimental entanglement chain length, $N_e = 96$ [24,178] is about three times larger than the simulation values. Thus, providing an explicit mapping between a real polymer and the bond fluctuation model makes it possible to give the scale factors D_{Rouse}, N_e that appear in Fig. 5.3 a specific quantitative meaning, as anticipated also by the comparison in Fig. 5.2.

The bond fluctuation model not only provides a good description of the diffusion of polymer chains as a whole, but also the internal dynamics of chains on length scales in between the coil size and the length of effective bonds. This is seen from an analysis of the normalized intermediate coherent scattering function S(q,t)/S(q,0) of single chains:

$$S(q,t) = \frac{1}{N}\sum_{i=1}^{N}\sum_{j=1}^{N}\langle \exp\{i\vec{q}\cdot[\vec{r}_i(t) - \vec{r}_j(0)]\}\rangle, \tag{5.6}$$

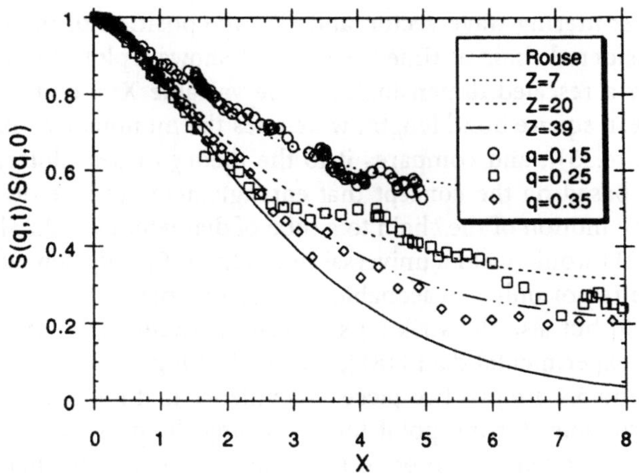

a

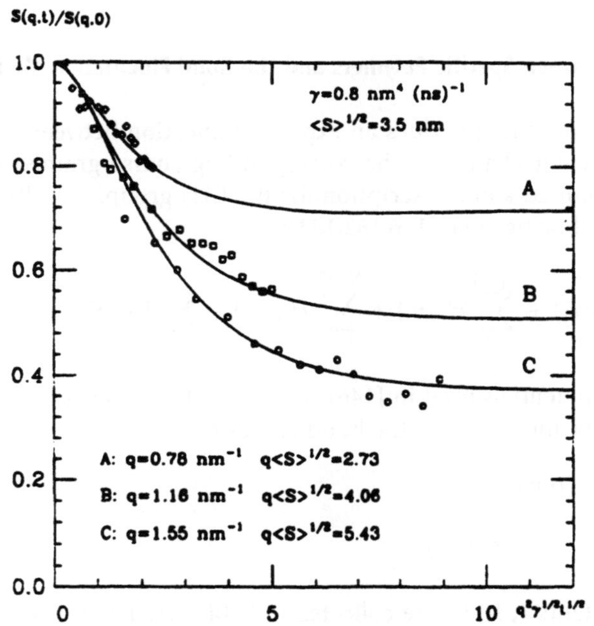

b

Fig. 5.4. **a** Scaled plot of the normalized intermediate scattering function S(q,t)/S(q,0) vs. the Rouse variable, $X = q^2 \langle \ell^2 \rangle \sqrt{wt}$, for the bond fluctuation model and three values of q (note that the unit of length here is the lattice spacing, and that $\langle \ell^2 \rangle$ and w denote the mean square bond length and the monomer relaxation rate, respectively). *Dashed curves* are results predicted from the theory of des Cloizeaux [179], where $Z = q^2\, d_T^2/6$ with a tube diameter $d_T = 44$, while the full curve is the Rouse model prediction [180]. All MC results refer to a chain length of N = 200, volume fraction $\Phi = 0.5$ of occupied sites. In this simulation, a bond angle potential $E(\theta) = \epsilon \cos\theta$ was also chosen, with $\epsilon/k_BT = 1$. From [42]. **b** Same as **a** but comparing the experimental data of Richter et al. [181] with the theory of des Cloizeaux [179], choosing a tube diameter (denoted as $\langle s \rangle^{1/2}$ in this figure) of 3.5 nm (experiment carried out using polydimethylsiloxane at room temperature).

where $\vec{q}$ is the scattering wave vector, and $\vec{r}_i(t)$ the position of the i'th monomer of the considered chain at time t. Figure 5.4 shows a plot of this function, choosing again a rescaled dimensionless time variable $X = q^2\langle\ell^2\rangle\sqrt{wt}$. [$\langle\ell^2\rangle$ denotes the mean square bond length; w denotes the monomer relaxation rate $\propto 1/\zeta$; see Eq. (4.25)], and compares it to the theory of des Cloizeaux [179]. This theory is based on the concept that entanglements can be described by constraining the motion of the chain to a tube of diameter d_T [22,23] while the Rouse model [21] would yield a universal function [180] only depending on X. The point here is not only a reasonable consistency between simulation [42] and theory [180] but also the striking similarity between simulation [42] and corresponding experimental data [181], included in Fig. 5.4b.

Figures 5.3 and 5.4 are just two pieces of evidence that the bond fluctuation model is a reasonable starting point for describing the properties of polymer melts. Thus the next step has to be to incorporate suitable information about the chemical structure and the energetics of specific polymers into the model.

5.3 The Mapping Between Specific Polymers and the Bond Fluctuation Model

The mapping procedure provides an explicit connection between an atomistic model of a polymer chain and the corresponding coarse-grained model. For PE, we use an united atom description for the CH_2 groups, resulting in a potential of the following type [146,182,183]:

$$\mathcal{H} = \sum_{j=1}^{N_p-1} \mathcal{H}_\ell(\ell_j) + \sum_{j=1}^{N_p-2} \mathcal{H}_\theta(\theta_j) + \sum_{j=2}^{N_p-2} \mathcal{H}_\phi(\phi_j) + \sum_{i\neq j} \mathcal{H}_{LJ}(\vec{r}_{ij}). \tag{5.7}$$

We choose the potentials used in [146] with a fixed bond length ℓ and harmonic potentials for the cosine of the bond angles θ_j:

$$\ell_j = \ell = \text{const} \tag{5.8}$$

$$\mathcal{H}_\theta(\theta_j) = \frac{f_\theta}{2}(\cos\theta - \cos\theta_o)^2, \tag{5.9}$$

where the constants ℓ, f_θ, θ_o are collected in Table 5.1. The torsional potential is described by:

$$\mathcal{H}_\phi(\phi_j) = f_\phi \sum_{k=0}^{5} a_k \cos^k \phi_j, \tag{5.10}$$

with the constants f_ϕ, $a_1, \ldots, a_5$ also quoted in Table 5.1 ($a_o \equiv 1$). Finally, the nonbonded interaction is expressed in terms of the Lennard-Jones parameters ϵ and σ (Table 5.1):

$$\mathcal{H}_{LJ}(\vec{r}_{ij}) = 4\epsilon[(\sigma/r_{ij})^{12} - (\sigma/r_{ij})^6], \tag{5.11}$$

Table 5.1 Parameters of the united atom force field for polyethylene used as the atomistic input for the coarse-graining procedure. The Lennard-Jones parameters pertain to CH_2-group interaction, since chain ends were not considered in the coarse-graining.

ℓ_{CC}	θ_0	f_θ	ϵ	σ	
1.53 Å	110°	$120\,\frac{\text{kcal}}{\text{mol}}$	$0.09344\,\frac{\text{kcal}}{\text{mol}}$	3.182 Å	
f_ϕ	a_1	a_2	a_3	a_4	a_5
$1.736\,\frac{\text{kcal}}{\text{mol}}$	$2.592\,\frac{\text{kcal}}{\text{mol}}$	$0.44\,\frac{\text{kcal}}{\text{mol}}$	$-4.032\,\frac{\text{kcal}}{\text{mol}}$	0	0

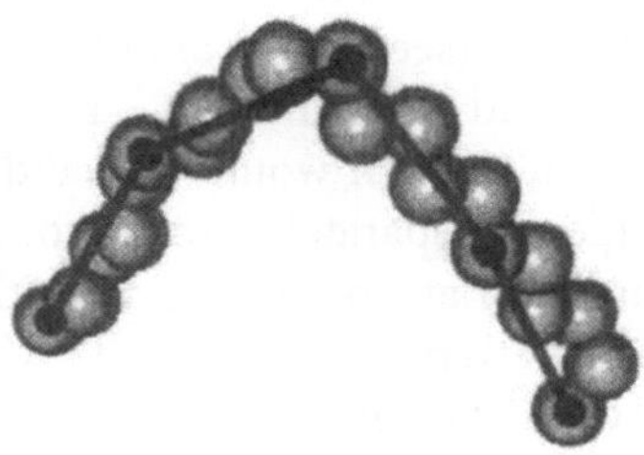

Fig. 5.5. Illustration of the coarse-graining procedure for a united atom chain. The chain is a segment of PE at 509 K from molecular dynamics simulations with the united atom model [Eqs. (5.7)–(5.11)]. One coarse-grained bond represents the end-to-end distance of n = 5 consecutive united atom bonds. From [32]

$r_{ij} = |\vec{r}_{ij}|$ being the distance between monomers i and j. While the potentials $\mathcal{H}_\ell$, $\mathcal{H}_\theta$ and $\mathcal{H}_\phi$ result from quantum-chemical calculations, the choice of $\mathcal{H}_{LJ}$ in Eq. (5.11) is phenomenological.

Using the potentials Eqs. (5.7)–(5.11), short single chains are simulated by simple sampling-type MC methods, as described in detail by Baschnagel et al. [27,29]. As discussed [29], a cutoff in the Lennard Jones interaction along the chain must be introduced in order for an isolated chain to show asymptotic Gaussian behavior. It is important to include the interaction of each unit along the chain with its fourth-nearest neighbor, to represent the pentane effect, while further distant neighbor interactions are eliminated [29,32]. A number of coarse-grained bonds m is then defined by connecting monomers n' repeat units apart (Fig. 5.5). Now both the distribution of the lengths L of these effective bonds and the distribution of the angle Θ between two consecutive coarse-grained bonds can be calculated. Figure 5.6 gives an example for $n' = 10$ and T = 509 K (note that for this temperature detailed atomistic molecular dynamics simulation of melts are also available, see Fig. 5.2a).

We now attempt to map distributions such as those shown in Fig. 5.6 over a wide range of temperatures (actually a range from T = 250 to 800 K was chosen [32]) by the bond fluctuation model, by equating m subsequent effective bonds of the bond fluctuation model to the n' chemical bonds of the atomistic model, and choosing the following temperature-dependent effective potentials for the lengths ℓ of the effective bonds and the angles ϑ between them (Note that ϑ is the angle between consecutive effective bands in the bond fluctuation model and Θ denotes the angle between consecutive coarse-grained bands in the band fluctuation model as well as in the atomistic model.):

$$U(\ell) = u_o(\ell - \ell_o)^2 + (1/T - \langle 1/T \rangle) u_1 (\ell - \ell_1)^2, \tag{5.12}$$

$$V(\vartheta) = v_o(\cos\vartheta - c_o)^2 + (1/T - \langle 1/T \rangle) v_1 (\cos\vartheta - c_1)^2, \tag{5.13}$$

where $\langle 1/T \rangle$ denotes an average over the temperatures where input information from the atomistic calculation (such as shown in Fig. 5.6) is used for the nonlinear fitting procedure, from which the 8 parameters u_o, u_1, ℓ_o, ℓ_1, v_o, v_1, c_o, c_1 and the time constant τ_o (see below) are obtained [30].

The proper choice of n' and m in this $n' \leftrightarrow m$ mapping is subtle [30,32], and there is no unique result which would be valid for all polymers; rather each case must be considered separately. On the one hand, it is desirable to choose these numbers as small as possible, to keep the effort manageable; on the other hand, n' has to be large enough so as not to impose too specific structural effects which could not be matched by the bond fluctuation model at all. In other words, n' has to be large enough, such that smooth distributions, as shown in Fig. 5.6, are actually obtained. On the other hand, the chemical structure of the polymer and the fact that a reasonable volume fraction of occupied sites (such as $\Phi = 0.5$) must correspond to melt densities impose further constraints (e. g., m effective bonds with bond angles $\vartheta = 0$ must reproduce the length of a piece of n' units of an all-trans PE chain, etc.). Considerations of this type fix $n' = 10$, $m = 2$ as the optimal choice, i. e. $n = n'/m = 5$ C–C bonds map into one effective bond, as anticipated in Fig. 5.5. The unit of length is then fixed so that this value yields the experimental density of PE at room temperature, which results in one lattice spacing corresponding to 2.03 Å [32].

We now discuss the translation of the MC "time-step" into physical time units. It is desirable to map the mobility of the lattice model (due to jumps of the effective monomers) onto the average jump rate of the torsional degrees of freedom, since these motions dominate the relaxation of the overall configuration of the chain. This means that we must allow for a temperature-dependent time unit $\tau_{MC}(T)$ which one attempted MCS per monomer corresponds to, via the formula:

$$\frac{1}{\tau_{MC}(T)} \left\langle \min\left[1, \exp\left(\frac{-\Delta\mathcal{H}}{k_B T}\right)\right] \right\rangle = \frac{1}{\tau_o} \frac{1}{N_{tor}} \sum_{tor} A_{tor}(T), \tag{5.14}$$

where $\Delta\mathcal{H}$ is the energy change in the MC move (to be calculated from Eqs. (5.12) and (5.13)), $1/\tau_o$ is the attempt frequency of torsional jumps in the atomistic model, N_{tor} is the number of torsions in the coarse-grained unit,

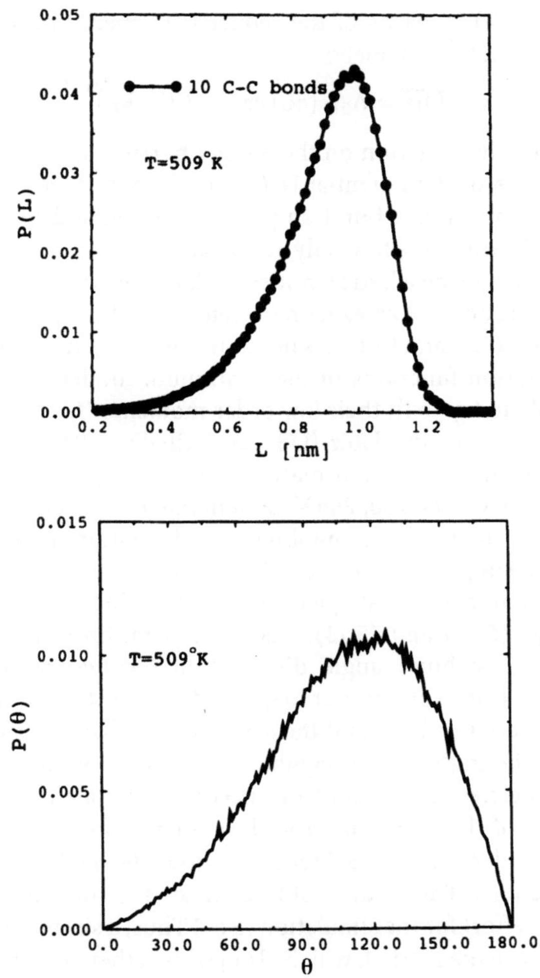

Fig. 5.6. Probability distribution of **a** the coarse-grained bond length L and **b** the bond angle Θ between two such coarse-grained bonds, choosing $n' = 10$ C–C bonds. From [32]

and A_{tor} is the average activated jump probability for the selected torsional degree of freedom. We require

$$\tau_{MC}(T) = A_{BFL}(\infty)\tau_o \exp(\Delta E_{min}/k_B T), \tag{5.15}$$

where ΔE_{min} is the smallest energy barrier in the torsional potential (Fig. 1.2b), and $A_{BFL}(\infty)$ is the acceptance of the MC moves in the athermal ($T = \infty$) bond fluctuation model. Writing $A_{tor}(T)$ in terms of an average energy barrier $\langle \Delta E \rangle$ defined as $\frac{1}{N_{tor}} \sum_{tor} A_{tor}(T) =: \exp[-\langle \Delta E \rangle / k_B T]$, Eq. (5.14) yields a further mapping condition for the Hamiltonian $\mathcal{H}$ (which contains the potentials $U(\ell)$, $V(\vartheta)$) in terms of the reduced (temperature-dependent) barrier $\langle W\rangle \equiv \langle \Delta E\rangle - \Delta E_{min}$, namely:

$$\langle \min[1, \exp(-\Delta\mathcal{H}/k_BT)]\rangle = A_{BFL}(\infty)\exp(-\langle W\rangle/k_BT). \tag{5.16}$$

Note that including information on the energy barriers into the mapping that fixes the parameters of the potential $U(\ell)$, $V(\vartheta)$ inplies that not all the many choices of bond lengths and bond angles can correspond to minima of the terminal potential, which is physically reasonable.

Now it is clear that using a lattice model, such as the bond fluctuation model, only an approximate and not an exact representation of the corresponding continuum model, Eqs. (5.7) and (5.11), is possible. Therefore no attempt is made to match the distribution functions of the continuum model (such as shown in Fig. 5.6) fully, only to try to fit their low order moments $\langle L\rangle$, $\langle L^2\rangle$, $\langle\Theta\rangle$, $\langle\Theta^2\rangle$. In addition we use a mixed correlator $\langle L\Theta\rangle$, and, finally, $\langle W\rangle$ is used in order to account for Eq. (5.16). All this information (for the temperatures T = 250, 300, 350, 400, 450, 509, 550, 600, 700, 800 K) is generated, see Table I of [32] for details, and used as input for a nonlinear optimization procedure [30,32]. Figure 5.7 shows a simplified flow chart illustrating this step. In order to calculate the average of an appropriate piece of the bond fluctuation chain from its Hamiltonian, Eqs. (5.12) and (5.13), exact enumeration techniques are used [30]. Considering the bond angle distribution for two pieces comprising $n' = 10$ C–C bonds (Fig. 5.6b), in our mapping four effective bonds of the bond fluctuation model are involved, and the same is true if one considers the jump rate because then the angles at the neighboring effective monomers also change. Even taking symmetries into account, one needs to generate 1747390 configurations for one step of this iteration procedure for this four-bond scheme. The total iteration considered in Fig. 5.7 requires about 100 h CPU-time on IBM RS 6000/370 workstations. This effort just to generate the starting Hamiltonian is comparable to the effort for the simulation itself! Thus, a simplified (but approximate) two-bond scheme is tried, which also yields rather reasonable results.

Figure 5.8 gives some examples to illustrate the extent to which the chemical input is matched. In view of the approximate character that a mapping to a lattice model necessarily involves, we consider the agreement satisfactory.

At this point, we will comment on how this procedure generalizes to other polymers. The other case that was considered by us [28,30,32,175,176] was concerned with bisphenol-A-polycarbonate (BPA-PC) (cf. Fig. 5.1). While for PE we had a correspondence that five chemical repeat units correspond to one effective bond of the bond fluctuation model, for BPA-PC the mapping ratio was inverse – one chemical repeat unit was mapped onto three effective bonds! One must consider, however, the very different sizes of the chemical repeat units: while for PE this is a single CH_2 group, in BPA-PC the repeat unit involves 12 C–C or C–O bonds along the backbone, and the end-to-end distance of the repeat unit is of the order of 10 Å. Thus in this case also one effective bond corresponds to a group of four successive covalent bonds along the backbone of the chain, and a lattice unit corresponds to about 2.03 Å [175].

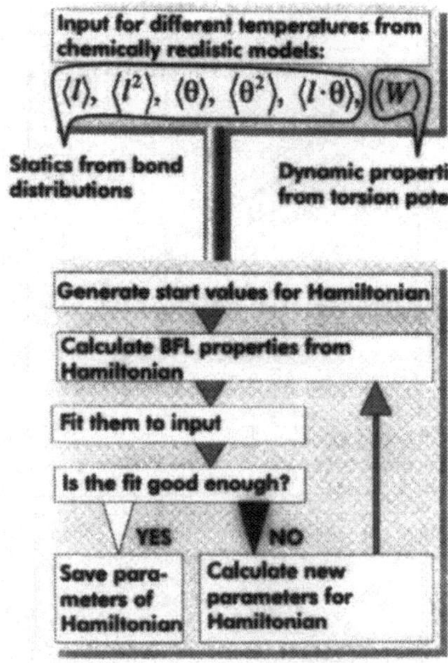

Fig. 5.7. Simplified schematic flow chart for the optimization of the parameters of the bond length and bond angle potentials. The input parameters from the chemically realistic model are the moments $\langle L\rangle$, $\langle L^2\rangle$, $\langle\Theta\rangle$, $\langle\Theta^2\rangle$, $\langle L\Theta\rangle$ taken from the bond length and bond angle distributions, and the reduced effective barrier $\langle W\rangle$ from the torsion potentials. From Tries [184]

From these considerations it is clear that the mapping procedure cannot be carried out for every polymer in precisely the same way, and it cannot be done "blindly". Every chemical monomer must be carefully considered for its structure and potentials and which mapping ratios n'/m are possible without contradictions that are physically sensible. The answer to this problem is not necessarily unique, and if several mapping ratios are possible, it may be best to try them all out - no general rule for choosing an optimal mapping ratio has yet been discovered.

5.4 Results of the Mapping Procedure

5.4.1 *Bisphenol-A-polycarbonate*

We now ask how well does the bond fluctuation model with these bond lengths and bond angle potentials reproduce the properties of real polymer melts quantitatively. First of all, it must be admitted that the model yields a qualitatively reasonable picture of the amorphous structure, as exemplified by

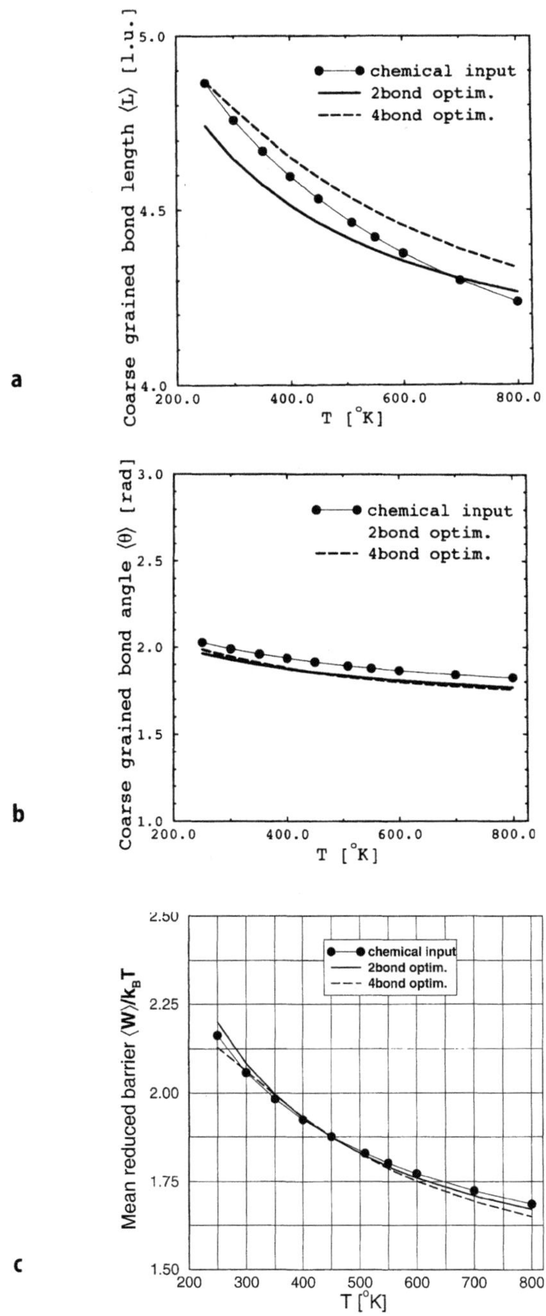

Fig. 5.8. Comparison of chemically realistic input for the optimization procedure and results for the corresponding lattice model, using the two-bond and four-bond optimization procedure: **a** mean length $\langle L \rangle$ (in lattice units), **b** mean anglē $\langle \Theta \rangle$, and **c** the mean reduced barrier $\langle W \rangle / k_B T$. From [32]

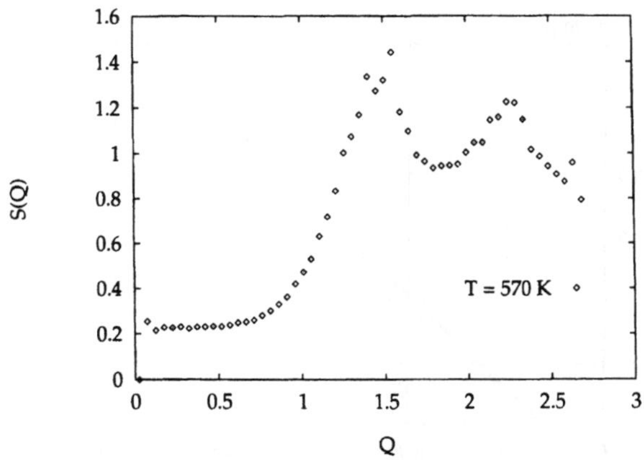

Fig. 5.9. Plot of the static collective structure factor S(q) for BPA-PC at T = 570 K, as obtained from mapping to the bond fluctuation model. From [175]

the static structure factor of the melt [175], Fig. 5.9, but it is somewhat unspecific with respect to the quantitative details. The structure factor of BPA-PC has its diffuse peak, the "amorphous halo", almost at the correct position, at about 1.4 $Å^{-1}$, and the decrease in S(q) towards small q is also reasonable. However, a quantitative fit of the experimentally measured structure factor [185] is clearly not achieved: for large wavenumbers q this is in principle also impossible; 3 $Å^{-1}$ roughly correspond to a lattice spacing a ($q \approx 2\pi/a$ with $a \approx 2.03$ Å), on such small length scales the lattice model is simply too coarse! What is more disturbing, therefore, is the fact that on the mesoscopic length scale also not all details are reproduced reliably: experimentally one finds in the structure factor S(q) a small side peak in the range $0.5 < q < 1$ $Å^{-1}$, that is not seen in the simulation (Fig. 5.9).

A key idea, in principle suitable to improve the description of the local structure, is "inverse mapping" [43,44,186]: the mapping step shown in Fig. 5.1 is now carried out in the opposite direction, using the equilibrated configuration of the coarse-grained model of the undercooled polymer melt as a starting point, replacing effective bonds by their corresponding chemically realistic units, which need to be relaxed with respect to their local orientation and internal conformation, of course, in order to minimize the energy. Since in practice this has not yet been tried – although a corresponding approach with a coarse-grained off-lattice bead-spring model yielded very encouraging results [43,44], see Sect. 6.3, – we shall not discuss this approach further.

In spite of the problems associated with the static structure, the coarse-grained model for BPA-PC did reproduce the glass transition of this material rather well: the self-diffusion constant of the chains follows the Vogel-Fulcher law [187] rather nicely (Fig. 5.10),

$$D = D_\infty \exp[-E_{VF}/(T - T_{VF})] \tag{5.17}$$

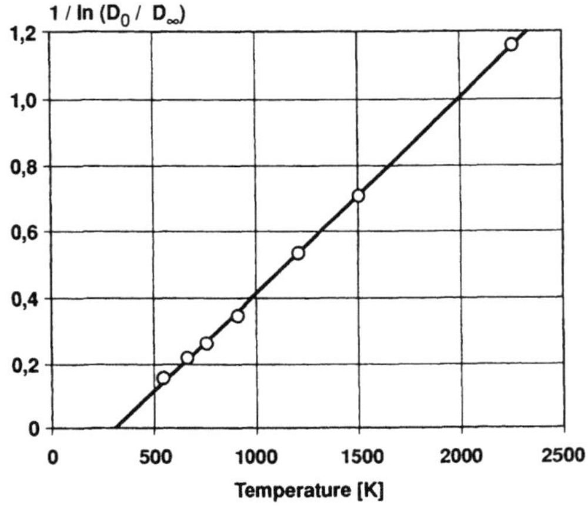

Fig. 5.10. Plot of the inverse logarithm of the self-diffusion constant of BPA-PC, for a length N = 20 of the coarse-grained chains, vs. temperature. *Straight line* indicates the Vogel-Fulcher [187] fit. From [28]

where D_∞, E_{VF} and T_{VF} are the diffusion coefficient at $T = \infty$, the Vogel-Fulcher activation energy and the Vogel-Fulcher temperature, respectively. T_{VF} should lie somewhat below the actual glass transition temperature. Plotting $1/\ln(D/D_\infty)$ vs. temperature, Eq. (5.17) suggests a straight line, $-(T-T_{VF})/E$, which intersects the abscissa at $T = T_{VF}$. Figure 5.10 shows the success of this analysis; $T_{VF} = 332$K is indeed close to the experimentally observed oligomer Vogel-Fulcher temperature $T_{VF}^{exp} = 322$K [186].

Unfortunately, it turns out that the good agreement between simulation and experiment is to some extent fortuitous - attempts aimed at reproducing differences between different polycarbonates (such as TMC-PC and BPA-PC) were less successful [186]. It is clear that the complexity of the chemical structure (Fig. 5.1) makes it very difficult to pin down the precise reasons for the successes and failures of the mapping procedure for this polymer.

5.4.2 *Polyethylene*

At this point we return to the polymer which is simplest with respect to its chemical structure, namely polyethylene (PE). In addition, for this polymer, the experimental database is much more complete, and also simulations of chemically realistic models, such as those described by Eqs. (5.7)–(5.11), are possible at high temperatures (Fig. 5.2a). Thus the prospects are very good that more can be learnt about the merits, as well as the limitations, of this modeling approach.

One decisive test is how faithfully the geometrical characteristics of single chains in the melt can be reproduced. Figure 5.11 shows that the characteristic ratio $C_N = \langle R^2 \rangle / N_p \ell_{CC}^2$ has a temperature dependence that is very similar to

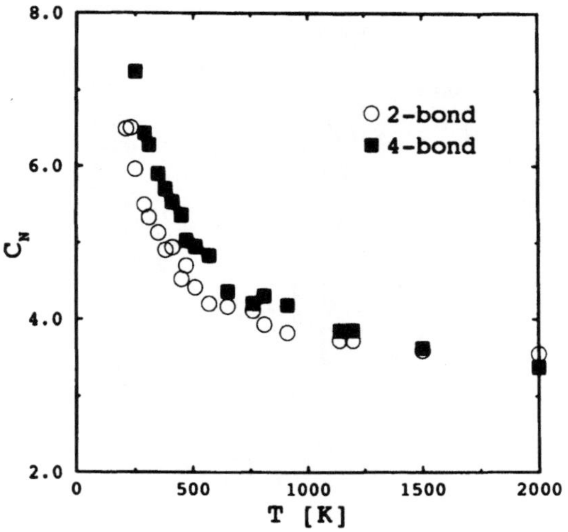

Fig. 5.11. Characteristic ratio of PE plotted vs. temperature, for N = 20 effective monomers. Both versions of the mapping, using the exact four-bond procedure and the approximate two-bond procedure, are included. From [32]

experiment. It is also apparent that C_N is systematically larger (and closer to experiment) for the four-bond mapping rather than the two-bond mapping.

It is, of course, important to ensure that the structure of the system stays fluid or amorphous throughout the considered temperature range, and that neither crystallization nor a transition to a liquid-crystalline order has occurred. Snapshot pictures of the configurations (Fig. 5.12) demonstrate clearly that no such ordering has taken place; instead, the characteristic disorder of interpenetrating and entangled polymer coils are observed. While at low temperatures there is clearly some chain stretching on the local scale, and the persistence length as well as the characteristic ratio C_N have clearly increased at low temperatures (see Fig. 5.11), the melt remains disordered.

The main motivation for our mesoscopic approach was to equilibrate supercooled polymer melts at as low a temperature as possible. For this purpose, it is advisable to use the "slithering snake" algorithm [188,189] instead of the random hopping algorithm that was described above. While the random hopping algorithm roughly corresponds to the real dynamics (jumps between minima of the torsional potential), the "slithering snake"-algorithm, where one moves an effective bond from one chain end to the other end in one step, does not correspond to any physical dynamics. However, this algorithm is faster by a factor of 100 to 1000 (Fig. 5.13), and in order to equilibrate the model properly and calculate static properties, it is a perfectly valid MC algorithm. It can also be used to generate an ensemble of well-equilibrated initial configurations which are then used for runs with the random hopping algorithm, which then yield information on dynamic properties at that temperature (self-diffusion constant, mean square displacements, relaxation times,

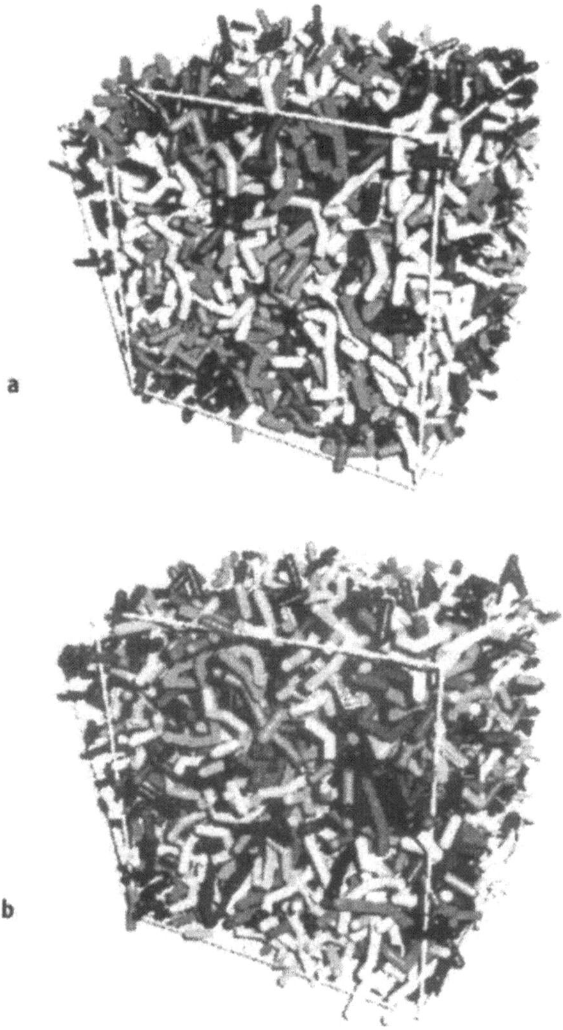

Fig. 5.12. Snapshot pictures of configurations of the effective bonds for PE melts at T = 2000 K (top) and at T = 210 K (bottom). Effective monomers are not shown, and effective bonds are shown as cylinders of different brightness. The edges of the cubic simulation box are also highlighted. Note that periodic boundary conditions are applied throughout. From [184]

etc.). The possibility to implement such unphysical global moves, in which phase space is transversed much faster, has no counterpart in molecular dynamics simulations and is an important technical advantage of MC methods.

We now discuss in more detail the time unit, $\tau_{MC}(T)$ [Eq. (5.15)], for the dynamic MC simulation. Remember that the Metropolis algorithm [7,12] only involves a transition probability, and in converting that into a transition prob-

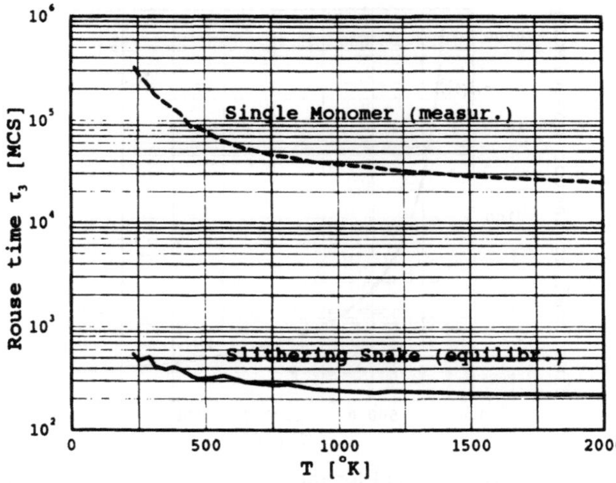

Fig. 5.13. Relaxation time τ_3 plotted vs. temperature for the coarse-grained model of PE with N = 20, using the random hopping algorithm (*upper set of data*) or the slithering snake algorithm (*lower set of data*), respectively. The time τ_3 is of the same order as the Rouse relaxation time of the chains, and is defined in terms of a crossing criterion for the mean-square displacements [41], $g_3(t = \tau_3) = g_2(t = \tau_3)$ [See Eqs. (5.2) and 5.3)]. From [32]

ability per unit time there is an inherent arbitrariness which can only be eliminated by additional considerations based on the actual dynamics that one wishes to model. We have included in the time unit the factor $\exp(\Delta E_{min}/k_BT)$, where ΔE_{min} is the smallest barrier of the torsional potential. The reasoning behind this choice is that, for a free polymer chain, conformational changes do occur with a rate proportional to $\exp(\Delta E_{min}/k_BT)$, and it is precisely these jumps which lead to conformational changes, that can be modelled by the jumps of the effective monomers that change the lengths of the effective bonds and angles between them. The energy difference of the minima of the torsional potential (Fig. 1.2b) then enters the effective potentials of the MC model: In the transition probability $P = \min\{1, \exp(-\Delta\mathcal{H}/ k_BT)\}$, cf. Eq. (5.14), only the energy of the initial and final state enter, and not the energy barrier between them. Therefore this barrier must be incorporated in the translation factor from the time unit of the MC process (1 MCS per monomer) to physical times (measured in seconds).

It only remains to specify the time constant, τ_o [Eqs. (5.14) and (5.15)], related inversely to the attempt frequency with which the monomers attempt to cross barriers in the torsional potential (Fig. 1.2b). We have not attempted to calculate this time constant from first principles, but rather fixed it by comparison to experiment on chain self-diffusion at T = 450 K [178]. This yields $\tau_o \approx 1/50$ picoseconds. This small number can be understood from the fact that because of the potentials, Eqs. (5.12) and (5.13), at T = 450 K only a few percent of the attempted hops of the effective monomers are successful: the time constant for successful hops is of the order of 1 ps. These considerations

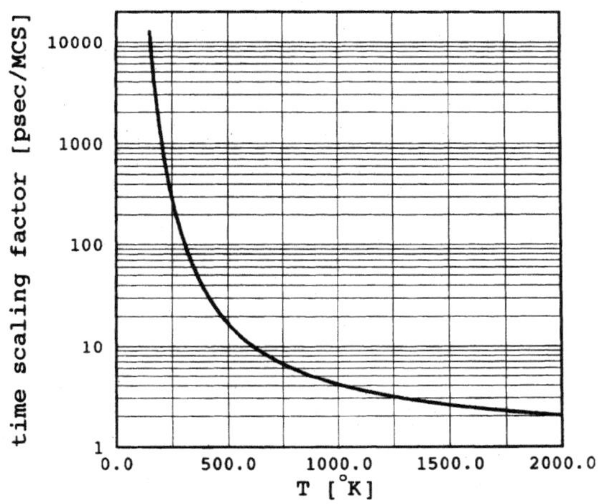

Fig. 5.14. Time rescaling factor $\tau_{MC}(T)$ for the coarse-grained MC model for PE plotted vs. temperature. From [32]

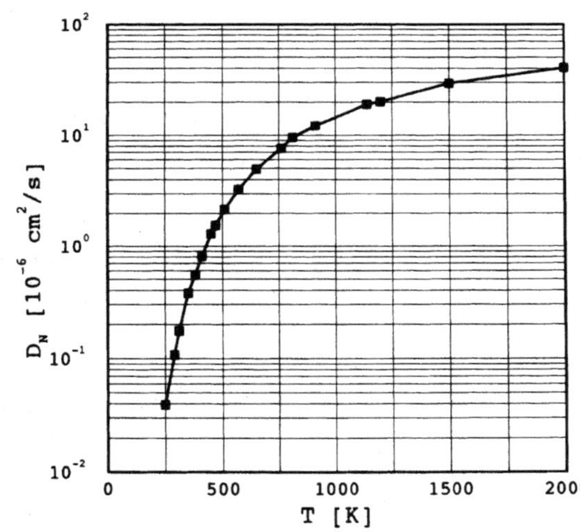

Fig. 5.15. Self-diffusion constant for PE chains (C_{100}) plotted vs. temperature, as predicted from the coarse-grained bond fluctuation model. From [32].

then yield the time rescaling factor $\tau_{MC}(T)$ as plotted in Fig. 5.14. By such methods it is then possible to observe the self-diffusion constant over about four decades (Fig. 5.15). In order to obtain the viscosity η as well, we rely on the Rouse model [21,22], using the formula:

$$\eta = \frac{\Phi}{8a^3} \frac{\langle R_g^2 \rangle}{6} \frac{k_B T}{N D_N} \text{ [Poise]}, \tag{5.18}$$

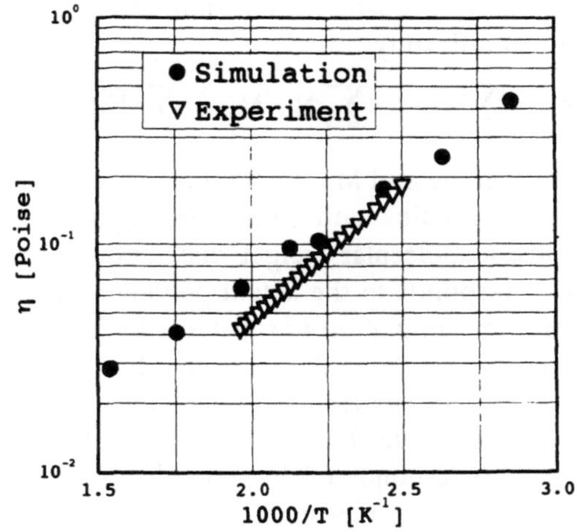

Fig. 5.16. Logarithm of the viscosity of PE melts (C_{100}), plotted vs. inverse temperature. *Full dots* are the predictions based on Fig. 5.15 using Eq. (5.18), while *open triangles* are experimental data of Pearson et al. [190]. From [32].

where $\Phi/8a^3$ is the density of effective monomers (volume fraction of occupied sites $\Phi = 0.5$, lattice spacing $a = 2.03$ Å). The result of this prediction is compared with the experimental data of Pearson et al. [190] in Fig. 5.16. Since there are no adjustable parameters whatsoever, we consider the reasonable agreement between simulation and experiment seen in Fig. 5.16 as significant. The fact that the experimental data vary more sharply than the simulation may be due to the fact that the simulation was carried out at constant density, while the experiment was done at constant pressure. Clearly, one of the weakest points of the present modelling based on the bond fluctuation model is that attractive intermolecular interactions have been omitted, and hence a reasonable equation of state (which would allow simulation of constant pressure data by varying the volume fraction Φ as a function of temperature such that the pressure in the simulation carried out for parameter combinations T, $\Phi(T)$ is approximately the same) has not yet been obtained. So far, the feasibility of introducing suitable attractive interactions into the bond fluctuation description of the glass transition has been demonstrated only for a schematic model [191], which has a two-level bond length potential only. Further research in this direction should be promising.

5.5
First Steps To Simulate Polydisperse Polymer Melts

All the simulations described so far refer to strictly monodisperse polymer melts, i. e. all polymer chains have strictly the same chain length N. For most experimental systems this is a drastic idealization, the polydispersity index p may differ from unity significantly:

$$p = M_w/M_n, \; M_n = \sum_i N_i M_i / \sum_i N_i, \; M_w = \sum_i N_i M_i^2 / \sum_i N_i M_i. \quad (5.19)$$

Here M_w is the weight average and M_n the number average molecular weight, assuming that there are N_i chains with molecular weight M_i in the melt. Ideally, in the simulations, a molecular weight distribution $P(N_i) = N_i/\sum_i N_i$ should be used that corresponds to the actual polymerization kinetics. However, this would require the simulation of substantially larger lattices than the $40 \times 40 \times 40$ lattices used here. Thus the treatment has been restricted to bi-disperse molecular weight distributions. Taking the mass unit equal to the mass of an effective monomer, we choose $M_1 = 20 - d$, $M_2 = 20 + 4d$, with $d = 0, 1, 2, \ldots, 10$ ($d = 0$ corresponds to the monodisperse case, where $p = 1$, while $d = 10$ yields $M_1 = 10$, $M_2 = 60$ and $p = 2$). Note that the number of chains must be chosen such that Φ is constant, i.e. the number of effective monomers in the system stays invariant, and hence $M_n = 20$ independent of d or p.

Assuming Gaussian statistics one expects $\langle R_i^2 \rangle \propto \langle \ell^2 \rangle M_i$ and hence the number average becomes:

$$\langle R^2 \rangle_n = \sum_i N_i \langle R_i^2 \rangle / \sum_i N_i \propto \langle \ell^2 \rangle \sum_i N_i M_i / \sum_i N_i = \langle \ell^2 \rangle M_n \quad (5.20)$$

so that $\langle R^2 \rangle_n$ is also independent of p or d. The same conclusion holds for the gyration radius. Figure 5.17 shows that the data are indeed consistent with this argument.

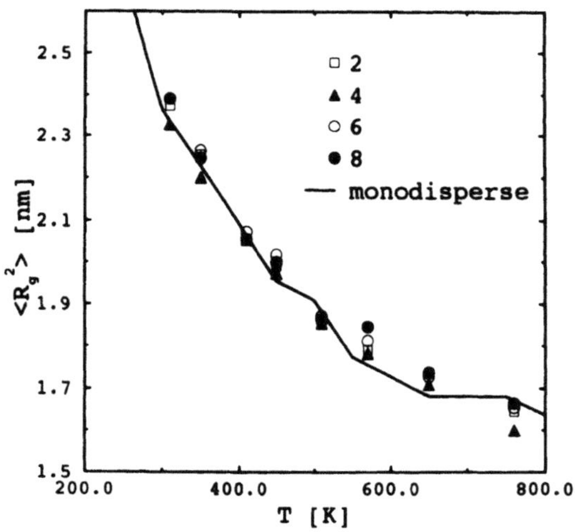

Fig. 5.17. Number average $\langle R_g^2 \rangle_n$ of the gyration radius of PE with $M_n = 20$ effective bonds plotted vs. temperature. Parameters of the symbols show the values of d. From [184].

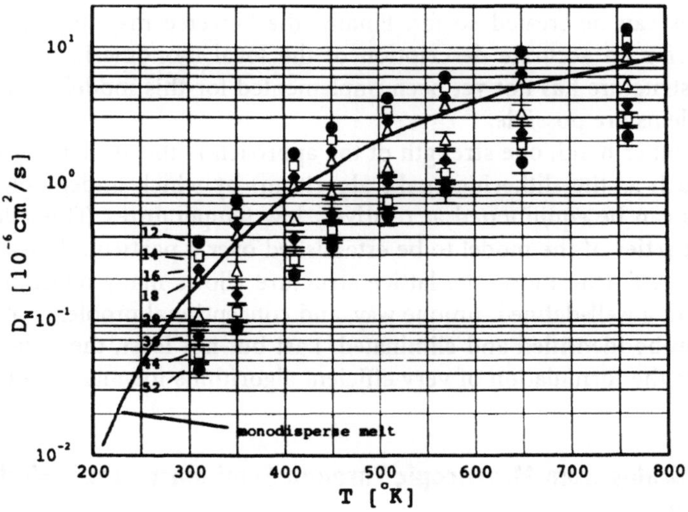

Fig. 5.18. Self-diffusion constants for a bidisperse (i.e. two different chain lengths) PE melt with $M_n = 20$ coarse-grained monomers. *Open triangles* are for d = 2, *filled diamonds* for d = 4, *open squares* for d = 6 and *filled circles* for d = 8. There are always two symbols of the same kind shown in the figure, since the bidisperse melt contains two species of different chain length. The numbers quoted in the figure correspond to these chain lengths for a given polydisparsity d. For instance, d = 8 corresponds to $M_1 = 12$ and $M_2 = 52$. From [184].

Dynamical quantities do behave differently, however. This is shown for the diffusion constant in Fig. 5.18, to demonstrate that there is a systematic trend that can be clearly observed in the simulations: the greater the polydispersity index p (or the parameter d) the larger the spread in the diffusion constant.

These results on bidisperse melts are clearly rather preliminary but, nevertheless, encouraging, since they show that the problem is now within reach of computer simulation analysis. Again this is a line of research which will require more efforts in the future.

5.6 Discussion

In this section, the state of the art of the lattice description of real polymers in terms of the bond fluctuation model augmented with bond length and bond angle potentials has been discussed. It has been shown that the approach has both merits and weaknesses.

Such weaknesses of the present implementation include the lack of an explicit inclusion of intermolecular forces other than excluded volume, resulting in a qualitatively inaccurate description of the equation of state. Another weakness is that the model shows lattice artefacts when dealing with problems of polymer crystallization or liquid-cristalline order: only rather flexible polymer chains can be treated so far. Finally, the "inverse mapping" procedure necessary for an accurate description of the local monomeric packing and the glass structure has not yet been implemented for this model, and unforeseen problems are possible.

On the other hand, one strength of the approach is the availability of algorithms (such as the slithering snake algorithm) by which undercooled polymer melts can be equilibrated at relatively low temperatures. This allows the static properties of the model to be established over a particularly wide parameter range. Furthermore, the lattice structure allows many questions to be answered in a well-defined, unique way, and conceptional problems of the approach can be identified and eliminated. Last but not least, the lattice structure allows the formulation of very efficient algorithms for many properties.

6 Bridging Scales from Microscopic Through Semi Macroscopic Models of Polymers

6.1 Introduction

As shown in the Introduction in Sect. 1, polymers can be characterized by a hierarchy of different length and time scales, and these scales, in particular the time scales, span an extremely wide range [15]. Figure 1.1 illustrates this and shows that the typical range that would have to be considered easily exceeds ten decades in time. On the microscopic level the relaxation properties and the dynamics are dominated by the local oscillations of bond angles and length. The typical simulation time constant is about 10^{-15} s, as indicated in the Figure 1.1. On the semimacroscopic level the behavior is dominated by the overall diffusion of the chains or the relaxation of the overall conformation of the objects [12,22,177]. These times, which depend on chain length and temperature, can easily reach seconds or, if the glass transition temperature is approached, even much longer times. To cover such a range within a conventional computer simulation is certainly almost impossible. On the other hand, it is important to relate the chemical structure of a system to the overall behavior of a material [26]. Thus, one of the long-standing challenges within the modeling of complex materials is to apply methods which allow the range from microscopic to semimacroscopic regimes to be covered [27–32]. Recently a method was developed which covers the mapping of polymers to a mesoscopic level as well as the reintroduction of the atomistic structure [43–44].

In this sense, similar to other contributions in this volume, we will attempt to bridge the gap from microscopic to mesoscopic and thereafter to the semimacroscopic [45] regime within a simulation scheme. Firstly, we will describe in detail a mapping procedure to go from a microscopic description of a polymer chain to a mesoscopic description which allows a fairly effective simulation procedure on a coarse-grained level [43]. The choice of three modifications of one polymer structure, namely polycarbonates, allows a detailed test of the sensitivity of this method. In order to check the quality of this approach the chemical details of the chains will then be reintroduced into the coarse- grained conformations and compared with neutron scattering results [44,185]. This will comprise the next two sections. Thereafter, first steps toward the next level of description are discussed. Starting from the conformations of polymer chains on the coarse-grained level, the chains are mapped onto an extended soft particle with only three internal degrees of freedom compared to the 0(3N) degrees of freedom of the whole object [45]. The aim of the present investigations is to provide a very general, as well as a very simple, approach to the simulation of specific polymers without losing the essential material properties. Unlike other earlier approaches [28,30], all methods will be within continuous space and not confined to a lattice structure. Though lattice simulations, especially the bond fluctuation approach [37–42], have been very successful, the advantages of continuous space simulations are striking. They allow for a more efficient packing and will in the future also be used for simulations with variable box size and shape.

6.2
Systematic Coarse-Graining Procedure

We will describe a systematic approach to renormalize the intrachain interactions towards a coarser level for three different modifications of polycarbonates. The advantage of examinig three modifications of the same polymer gives a first hint of the sensitivity of the method. The three modifications of the polycarbonate are BPA-PC, BPZ-PC, and TMC-PC. The structures are given in Fig. 6.1. Although the backbone sequence is the same they have re-

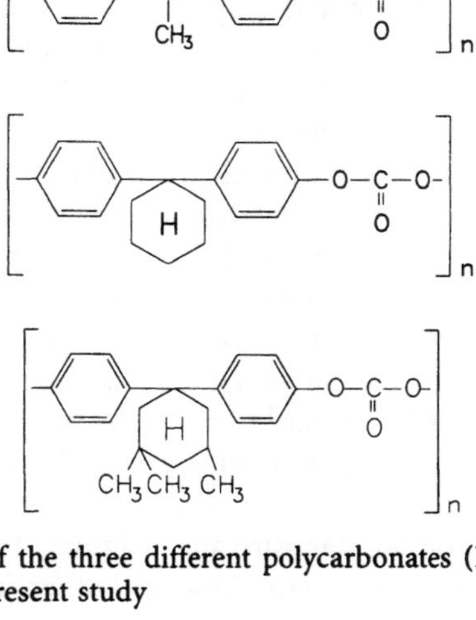

Fig. 6.1. Repeat units of the three different polycarbonates (BPA-PC, BPZ-PC, TMC-PC from top) used in the present study

markably different physical properties [186]. For the first two (BPA and BPZ), the glass transition temperature, T_G, is roughly the same ($T_G \approx 420\,K$), while the third one has a glass transition temperature which is approx. 80 to 100 K above the previous one, namely around 500 K. On the other hand, BPA-PC, and to some extent TMC-PC, are ductile while BPZ-PC is more brittle. This is also reflected in the difference in the generalized activation energy within the Vogel-Fulcher [see Eq. (6.4)] scheme. Here, again, BPA and TMC have roughly the same activation energy, while BPZ has a significantly higher activation energy. Although the general structure along the backbone is similar in all three cases, the glass transition properties and also the entanglement chain lengths N_e in the melt are significantly different. For BPA-PC an extremely short entanglement length namely $N_e = 7$ monomers only is reported. This length increases through $N_e = 9$ monomers (BPZ) to $N_e = 14$ monomers for TMC-PC. In particular the extremely short length for BPA-PC is not yet fully understood. Considering other well-studied polymers like polyethylene or PDMS (poly-dimethyl-siloxane), one would expect an increase by a factor of at least 5 to 10. Whether this is the result of a special nonuniversal behavior of the system on short length scales (resulting from the special form of the banana-shaped repeat units joined by almost pivot-like junctions) is beyond the scope of the present rewiew and will be studied in a future investigation.

The coarse-graining procedure will be explained in detail for BPA-PC, but is similar for the other two modifications. The aim is to develop a method which is parameter free and as simple as possible. In addition we will attempt to stay as close as possible to the chemical structure in order to then be able to reintroduce the chemical details without too many problems. Thus, the coarse-grained monomers have to be designed in a way that they can easily be identified with specific chemical groups of the polymer itself. Taking the chemical structure of the three different polycarbonates, a 2:1 mapping seems to be a first reasonable choice, as illustrated in Fig. 6.2. The resulting coarse-grained structure then only has four relevant internal degrees of freedom: the bond length between carbonate and isopropylidene group, two bond angles and the torsion angle, meaning that all the complicated intramolecular interactions and degrees of freedom are mapped onto these four intramolecular interactions. To arrive at these coarse-grained interactions from the microscopic model a number of fitting procedures could be envisaged. Here, however, we follow a different route. Since the coarse-grained potentials do not only have to include energetic aspects of the microscopic model but also entropic parts from the different possibilities of local conformations, we use intrachain distribution functions to construct the bonded potentials in the coarse-grained model. Knowing the potential functions of the detailed chemical system it is rather straight-forward to simulate, to a very high accuracy at a given temperature, the conformations of individual free random walks. The probability distribution functions of conformations of such a model system are then only dependent on temperature and originate from the bonded interactions along the backbone of a chain. Since no other interactions have been

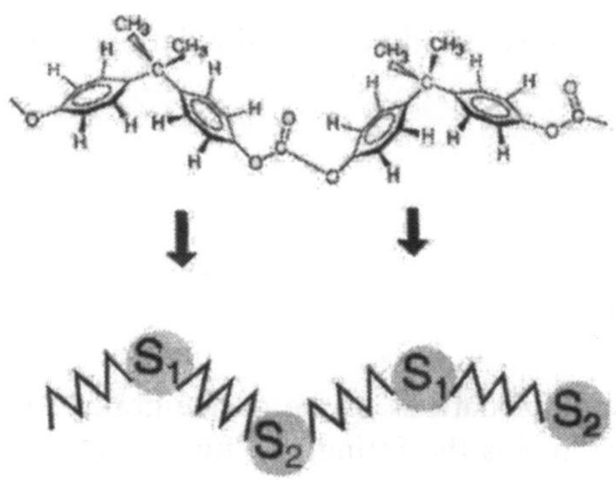

Fig. 6.2. Illustration of the mapping procedure for a 2 : 1 mapping where the repeat unit of a BPA-PC chain is replaced by two monomers of a generalized bead spring chain. The geometrical centers of the carbonate group and the geometrical center of the isopropylidene group, respectively, are mapped onto the centers of the new spherical beads. From [43]

taken into account, the distribution functions at any given temperature can be generated to a very high accuracy. The potentials for the microscopic models are derived from ab initio quantum chemistry calculations. Using this approach we can directly sample the probability distribution function $P(\ell, \alpha, \beta, \delta)$ for the coarse-grained model within the limits of single isolated random walks. ℓ is the bond length of the coarse-grained model, α the carbonate-isopropylidene-carbonate bond angle, β the isopropylidene-carbonate-isopropylidene bond angle and ϑ the torsion angle of the coarse-grained model, as sampled by the microscopic simulation. The coarse-grained distribution function is temperature dependent via the Boltzmann-weights of the different states of the microscopic model. The most crucial assumption now is, that the distribution function of the set of variables factorizes into independent distribution functions of the individual variables:

$$P(\ell, \alpha, \beta, \vartheta) = p(\ell)p(\alpha)p(\beta)p(\vartheta) \tag{6.1}$$

For polycarbonates this was checked to be a good approximation [43,195]. Since each distribution function P is determined at each temperature separately, the individual p values for different temperatures give the thermodynamic probabilities of a given state, which simply means that:

$$p(\ell) \propto \exp(-U(\ell))$$

$$p(\alpha) \propto \exp(-U(\alpha)) \dots \tag{6.2}$$

where U is a corresponding generalized potential for the chosen simulation temperature, already normalized to this temperature. This normalization allows the simulation temperature to be kept at $k_BT = 1$, which is of technical advantage for standard molecular dynamics simulations. Using Eq. (6.2) we obtain the forces:

$$F(\ell) = -\frac{d}{d\ell}\ln p(\ell)$$

$$F(\alpha) = -\frac{d}{d\alpha}\ln p(\alpha)\ldots \tag{6.3}$$

which give rise to the conformations of the coarse-grained model. This simple and direct approach avoids the fitting of a functional form of the coarse-grained potential functions to the microscopic parameters. The only fitting procedure which has to be implemented is a smoothing of the potentials in order to diminish the scattering due to statistical noise of the sampling. This is necessary for a proper derivation of the forces, which are simply tabulated. There is no need to determine the partition function explicitly since it shows up only as a constant in the potential and thus does not alter the resulting forces. Figure 6.3 gives a typical example for the angle. By means of this model we can now simulate dense polymer systems. The volume of the coarse-grained monomers, which are represented by effective hard spheres (repulsive Lennard-Jones), are adjusted to give the same van der Waals volume as in the experimental case (normalized to the simulation density). In the case of BPA-PC the carbonate group and the isopropylidene group are represented by spheres with radii of 3.02 and 3.11 Å, respectively. No further specific excluded volume interactions or any directional interaction is taken into account. The simulations are then performed as molecular dynamics simulations at constant volume. The simulation density is adjusted to the experimental mass density of the different polycarbonate systems. Besides this there is no freedom to adjust parameters in the model beyond the origi-

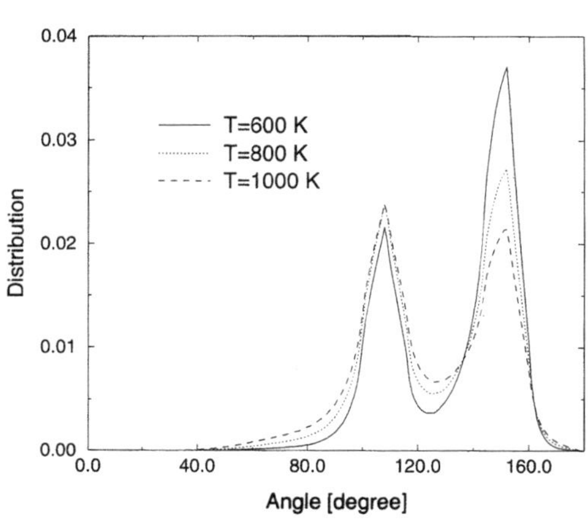

Fig. 6.3. Temperature-dependent distribution of the angle β ($\angle\ S_1S_2S_1$, cf. Fig. 6.2) for different temperatures [43]

nal setup of the procedure. If this procedure is able to reproduce the essential aspects of the different chemical species then the static structure which comes out of this simulation should compare well to the experimental systems. The dynamics as well as the statistic of a system can be reproduced by this numerical approach. A detailed discussion of the dynamic properties is given in [43].

While for the static properties a scaling of lengths is given from the mass density of a system, for the dynamic properties the scaling is only fixed up to a constant. This constant is determined by comparison of the viscosities of a simulated system and the experimental system. During the melt simulations the excluded volume interactions of the monomers are taken into account via the repulsive part of a Lennard-Jones interaction. Starting from a mass density of $\rho_{PC} = 1.05$ g/cm^3 (BPA-PC at 500 K) and the simulation number density of $\rho_{MD} = 0.85\sigma^{-3}$, we arrive at a length scaling of $1\sigma = 5.56$ Å for the present case (σ being the unit length of the Lennard-Jones potential). The simulation number density is the usual one from melt simulations of coarse-grained models. The number density in the simulation is scaled with the temperature dependency of the mass density. Thus the scaling factor stays constant for all simulated systems at all temperatures while the simulation temperature is kept at $k_B T = 1$ (see above). Simulations were performed by integrating Newton's equations of motion for a coarse-grained model with a velocity Verlet algorithm [9,192]. In order to keep the temperature constant, and to stabilize the algorithms, all particles are weakly coupled to a heat bath and background friction (standard procedure for melts [177]). For the present system the background friction is about 100 times weaker than the monomer-monomer friction, and therefore irrelevant for the results. The simulation systems typically comprised between 1000 and 10 000 model monomers of chains of 20 or 60 model monomers, resulting in systems between 500 and 5000 real chemical repeat units. For our cubic simulation box this means that systems of up to 125 Å^{3}can easily be considered. The resulting effective speed up compared with microscopic simulations is of the order of 10000. It turns out that the interchain interactions strongly modify the angular distribution functions compared with the isolated chain. This is important as it means that there is no simple random walk equivalent derived from isolated chains for the local chain conformation! To employ the static properties in more detail we later reintroduce the complete chemical structure. For the coarse-grained model we first check the dynamical properties as a function of temperature, especially as a function of approach to the glass transition temperature. As for polycarbonates, the properties of many materials, when approaching the glass transition temperature, are well described by the so-called Vogel-Fulcher behavior:

$$D = D_o \exp\left(-\frac{A_o}{T - T_{VF}}\right) \tag{6.4}$$

which is given here for the tracer diffusion constant D of the chains (cf. Sect. 5). D is the tracer diffusion constant of the polymers, given by ($\vec{r}_{CM}$ being the position of the center of mass of the chains):

$$D = \lim_{t\to\infty} \langle (\vec{r}_{CM} - \vec{r}_{CM}(t))^2 \rangle / 6t \tag{6.5}$$

A_o is a generalized activation energy and T_{VF} is the hypothetical zero temperature. T_{VF}, the so-called Vogel-Fulcher temperature, is typically about 80 °C below the calorimetric glass transition temperature. The prefactor D_o is a hypothetical high temperature diffusion constant which describes the system without any specific interaction. For the present situation D_o is easy to determine, because it simply corresponds to the freely jointed polymer melt with excluded volume where all the chemically dependent intramolecular interactions are set to 0. For this model extensive computer simulations are available and values of D_o precisely determined. Figure 6.4 gives a typical Vogel-Fulcher plot of the three different polycarbonate modifications. The results qualitatively match perfectly the experimental situation, namely that the Vogel-Fulcher temperature for TMC-PC is about 80–100 °C above the Vogel-Fulcher temperatures of BPZ-PC and BPA-PC while the generalized activation energy, which in Fig. 6.4 is the slope of the lines, is roughly the same for BPA and TMC while it is different for BPZ-PC. Even quantitatively the results are not that different from the typical experimental value, as Table 6.1 shows.

Within the Rouse model for polymer dynamics the viscosity of a melt can be calculated from the diffusion constant of the chains using the relation [22,29,30]:

$$\eta = \frac{\rho < R^2 >^2}{s^3(N-1)^2 < \ell^2 >} \frac{k_B T}{36D} \quad [Poise \frac{\tau}{sec}] \tag{6.6}$$

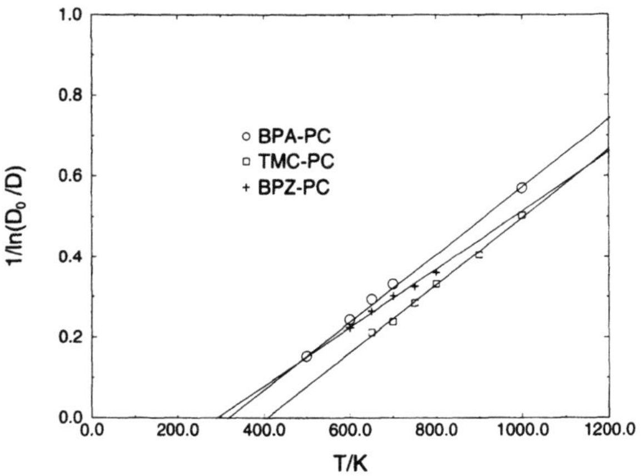

Fig. 6.4. Vogel-Fulcher plot of the chain diffusion constants D for the three different polycarbonate modifications, as indicated in the figure, for N = 20 model monomers

Table 6.1 Activation energies A_0 [K] (below) and Volgel-Fulcher temperatures T_{VF} [K] (above) for experiment and simulation. While the error bar for T_{VF} is fairly small, it is significant (≈30%) for the experimental determination of A_0 due to the large polydispersity of the typical commerical samples

T_{VF}	BPA-CP	TMC-PC	BPZ-PC
Simulation (N = 20)	233	407	292
Experiment	387	477	392
A_0	BPA-CP	TMC-PC	BPZ-PC
Simulation (N = 20)	1305	1363	1443
Experiment	1012	1073	1534

with $<R^2>$ being the end-to-end distance of the chain and s the length scaling factor between the experimental system and the model system, N the chain length, $\langle \ell \rangle^2$ the squared mean bond length and τ the simulation time unit within the Lennard-Jones framework. Using Eqs. (6.5) and (6.6) we can compare the model dynamics with the melt dynamics of the different species, since we have one case where the highest experimental temperature and the lowest simulation temperature coincide. This gives the relation

$$1\tau = 2.21 \times 10^{-10}\text{sec} \tag{6.7}$$

The simulation time step is typically $\delta t = 0.01\tau$. Equation (6.7), however, should only be used as a guide, since the experimental systems and the simulation systems comprise different chain lengths; in addition, the effect of polydispersity might alter this absolute scale by some prefactor. Compared with other molecular dynamics simulations of microscopic models the simulation time step is roughly three orders of magnitude larger. Taking the simplicity of the potentials and the short range nature of the interactions into account the resulting speed up is of the order of $\mathcal{O}(10^4)$.

6.3
Inverse Mapping from the Mesoscopic Back to the Microscopic Regime

There are various ways to check the quality of the resulting structures with respect to experiment. A typical check would be to compare the mean square end-to-end distance $<R^2>$ with results from scattering experiments. However, since the experimental samples are highly polydisperse, the results from scattering experiments are somewhat questionable [195]. Furthermore, a crucial check is the direct comparison of conformations of systems. In order to be able to compare the conformations resulting from simulations unanimously to experiment we reintroduce the chemical details into the coarse-grained chain. This is one of the reasons why it is so important to device a mapping procedure which stays close to the chemical structure of the objects. We have a one-to-one correspondence of the model monomers to the different parts of the chemical repeat unit of the chains. To do this at the present time we revert to the use of commercial packages, since details of the force fields are no longer of essential importance. In the future this will also be improved. To reintroduce the details we start out with a chemically detailed chain with the correct bond angles and bond lengths. Then the chain is placed in the system and rotated via the torsional degree of freedom along the path of the coarse-grained chain. The structure is optimized via a simple steepest descent algorithm. Using this approach all individual chains are mapped onto their chemically detailed chains individually. The achieved deviations between the minimized structure and the starting structure are as small as $\langle \Delta r^2 \rangle = 0.01$ Å^2. In a second step all the melt of the chemically detailed chains is relaxed locally after introducing van-der-Waals interactions. This local equilibration of course only produces motions over a very small distance (≈ 1 Å). Since for these resulting packing structures the polydispersity effects of the experimental melts should not be that important, we can now calculate the coherent structure function of our systems and compare them directly to neutron scattering data. The structure function S(q) is given by:

$$S(q) = |\sum_j b_j \exp(i\vec{q} \cdot \vec{r}_j)|^2 . \tag{6.8}$$

For this comparison all atoms of the systems are explicitly included with their corresponding scattering length b_j [193]. Figures 6.5 and 6.6 give two typical examples for BPA-PC. Figure 6.5 shows S(q) of a fully protonated system and compares the data with both neutron scattering and previous amorphous cell simulation. Figure 6.6 gives the results for the case of deuterated methyl groups.

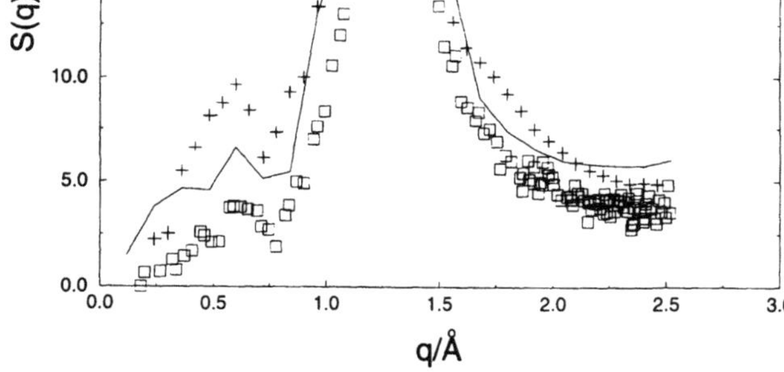

Fig. 6.5. Coherent structure function S(q) in absolute units in comparison to amorphous cell simulations [194] and neutron scattering data [185]

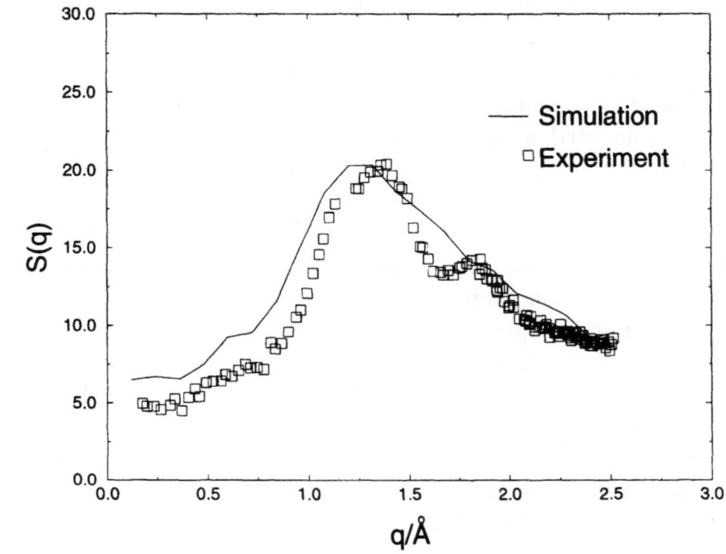

Fig. 6.6. Comparison of experiment and simulation for the same case as above (Fig. 6.5), but now for a system with deuterated methyl groups (experimental data from [185]

Unfortunately, for the deuterated systems, no amorphous cell data are available. Other considered structure functions show the same agreement to experiments while for TMC-PC the density fluctuations are greater than for BPA-PC which leads to smeared out peaks for the structure factor. The data are in excellent agreement with experiment. The partially deuterated examples actually show that the simulations are able to reproduce details of the scattering curves down to the smallest possible values of q compared with the box size. This is more crucial than the direct comparison of the average chain extensions since, as can be shown by simulations on the coarse-grained level, the interpretation of scattering data as $\langle R_G^2 \rangle$ is rather questionable for polydisperse systems [195]. Similar problems show up in a comparison of simulation data with dynamics in order to determine the entanglement length. It transpires, that the apparent small entanglement length for BPA-PC is most probably a result of a mixture of universal and nonuniversal effects [195]. Computer simulation and neutron scattering spin echo experiments typically deduce the entanglement length from short time data while rheology does this from long time data. Detailed comparison for polycarbonates would be very useful in order to understand these deviations from universal behavior somewhat better.

6.4
An Even Coarser View of Polymers

So far all the systems which have been considered still allow the one-to-one correspondence of a bead of a polymer to the repeat unit of a given chemical species. This still means that the number of degrees of freedom that have to be considered are proportional to the number of monomers of a given chain, causing enormous problems if large systemsneed to be simulated, i.e. not many monomers but many chains. In order to arrive at a situation where many chains can be simulated we go back to Fig. 1.1. There three levels of description were illustrated. The previous two sections discussed mapping from the microscopic to the mesoscopic regime and backwards. Now we want to consider another step, namely to try to map the chains from the mesoscopic system up to the semimicroscopic regime where we replace the chain by soft ellipsoidal particles which can overlap strongly in the melt [45]. For such a situation each chain is represented by a soft ellipsoid which varies in size and shape. Thus the whole chain is replaced by a particle with three internal degrees of freedom given by the free axes of the mass tensor of the chains. To do this we follow a philosophy which is very similar to the coarse-graining procedure discussed before. The idea is to separate the free energy of a chain into an intrachain part and an interchain part. Thus, for the total free energy F, the following relationship can be deduced:

$$F = \sum_{i=1}^{M} F^{i}_{\text{intra}} + F_{\text{inter}} \tag{6.9}$$

The first sum runs over all M chains of the system. First let us consider the intrapart of the free energy. In a melt the topologically allowed conformations of a selfavoiding walk are the same as in vacuum. The change from an extended coil state to a random walk state in a melt is a result of the interchain interactions but not of the intrachain distribution of the conformations. Thus, we characterize the intrachain contribution of free energy by the number of states which correspond to the given mass tensor of the selfavoiding walk. With $\underline{\underline{R}}$ being the inertia tensor of the chain with the eigenvalues R_1, R_2, R_3 (with $R_1 < R_2 < R_3$) we generate, similar to that described in the previous sections, a probability distribution of this mass tensor $\underline{\underline{R}}$, namely $P(\underline{\underline{R}})$. For each given $\underline{\underline{R}}$ there is a corresponding intrachain density distribution $\rho(\vec{r}, \underline{\underline{R}})$ which is also sampled. The averaging is carried out over all conformations with a given $\underline{\underline{R}}$. Taking into account that the topologically allowed conformation of individual chains in the melt and for the isolated chain are exactly the same, the intrachain contribution to the free energy from chain i is given simply by:

$$F^{i}_{\text{intra}} = -k_B T \ln P(\underline{\underline{R}}^{i}) \tag{6.10}$$

Now we assume that the inter and the intrapart are additive in a way that the inter chain interaction is given by the pairwise overlap of the ellipsoids of the different chain. Since each mass tensor corresponds to a density distribution we can describe the interchain free energy contribution of the pair ij:

$$F^{(ij)}_{\text{inter}} = \epsilon(N) \int \rho_i(r, \underline{\underline{R}}^{i}) \rho_j(r, \underline{\underline{R}} j) d^3 r \tag{6.11}$$

Here each of the two density distributions is centered on the center of mass of the corresponding chain, with the local axes oriented along the principle axes. This results in a total free energy:

$$F = -k_B T \sum_i \ln P(\underline{\underline{R}}^i) + \epsilon(N) \sum_{i \neq j} \int \rho_i(r, \underline{\underline{R}}^i) \rho_j(r, \underline{\underline{R}}^j) d^3 r \tag{6.12}$$

$\epsilon(N)$ is an adjustable parameter accounting for the binary excluded volume as well as the overlapping contribution of the probability distributions. For technical details refer to [45]. The systems we use to test this idea are simple coarse-grained polymer models, as they are frequently employed in the study of polymeric melts and networks. Extension to a more refined coarse-grained model, for example, polycarbonate is straightforward and is the subject of work in progress. Within this scheme it is easy to simulate on one DEC-Alpha processor polymeric melts of the order of several thousands of particles meaning chains, respectively. Typical systems consist, for instance, of 10 000 chains of N = 100 monomers. The simulation is a standard Metropolis Monte Carlo simulation procedure as described in the earlier chapters of this volume. The ellipsoidal particles are randomly distributed in the system starting from their self-avoiding walk distribution function. Then a Monte Carlo simulation is performed in such a way that the ellipsoids can move in space and can change their principal axes resulting not only in a shape deformation but also in a rotation of the objects. Figure 6.7 gives a typical evolution plot of the radius of gyration of the ellipsoids in our investigation as a function of Monte Carlo steps for different parameters of $\epsilon(N)$. As Fig. 6.7 shows the adjustment

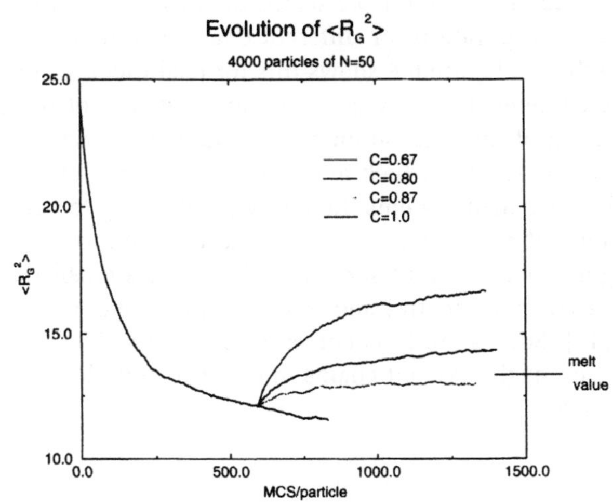

Fig. 6.7. Evolution of the sample averaged $\langle R_G^2 \rangle$ as a function of MC time. The initial value of $\varepsilon(N) = C = 1.0$ was changed to the values indicated after 600 MC steps. The indicated melt value corresponds to a comparable system with explicit chains with repulsive Lennard-Jones interactions and a number density of 0.85 σ^{-3} (from [45])

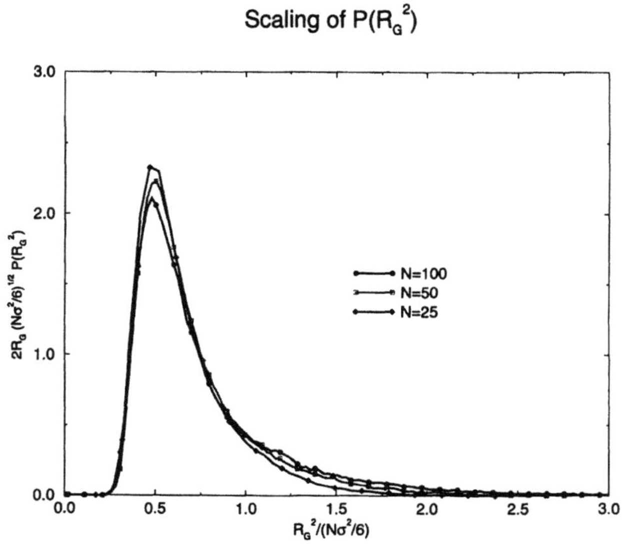

Fig. 6.8. Scaled distribution functions of $R_G^2(N)$ vs. the Gaussian-normalized value of R_G for chain lengths as indicated (from [45])

of ϵ allows a precise mapping of the ellipsoidal model onto the explicit chain model at a given density. Note that the distribution functions of the ellipsoids were generated from the selfavoiding walk simulations of the explicit chains. In order to show that the end-to-end distance of the ellipsoidal system in the melt agrees with the explicit chain simulation, and also that the chain statistics correspond to Gaussian statistics, we scaled the probability distribution function of the radius of gyration for different chain lengths within the random walks scaling scheme. Figure 6.8 shows this for chain lengths between 25 and 100. Again the agreement is very good. Various other control investigations, such as the scaling of the correlation hole, support this picture. Thus, we arrive at a simulation scheme which allows us to simulate almost macroscopically large polymeric melt systems. In a very similar way to earlier studies on phase separations of polymers, we can now also introduce an $\epsilon(N)$ which is able to distinguish two different species [45]. This has enabled the investigation of phase separation kinetics and morphology development of huge polymer samples [45]. Such a study is currently underway. The next step will be the reintroduction of the explicit chains in order to complete the scheme as it is given in Fig. 1.1.

6.5 Conclusions

Section 6 has considered various approaches to bridge the gap between the microscopic and the semimicroscopic regime for simulation of complex polymer materials. It is obvious that the described approaches are still only very first attempts. However, together with the experience described in other chapters in this volume, a number of different methods are now being developed. Also ways to generate coarse grained effective force fields are being discussed [56a]. Considering the progress in these algorithms as well as the improvement in the computer hardware, such methods will become increasingly important for complex macromolecular systems.

7 Application of Atomistic Modeling Techniques to Heterogeneous Polymer Solids

On a continuum level, the adjective "heterogeneous" describes a material composed of different substances or the same substance in different phases; its complement, "homogeneous", indicates that the material is composed of parts that are all of the same kind. Clearly, the border between heterogeneous and homogeneous depends on the length scale applied (e.g., on an Å scale any substance is heterogeneous). When multiple length scales are involved, the question of heterogeneity versus homogeneity have to be addressed. At distances exceeding ca. 100Å, the approaches of atomistic, meso-scale, and continuum modeling overlap. Developments in the modeling of composites have recently made the accurate modeling of heterogeneous continua possible and we describe here briefly the concepts involved. For details, the reader is directed to the original literature.

As for any modeling of continuum structures, the properties of the phases must be known for this kind of approach to work. Here, estimates obtained by atomistic methods of other techniques, described in the earlier chapters of this review, may be employed, or empirically known values may be used. It is hoped that the co-development of these continuum techniques and atomistic and coarse-grained approaches will lead to a seamless integration of the different techniques.

7.1 Direct Energy Minimization Versus the Finite Element Method

The modeling of complex solids has greatly advanced since the advent, around 1960, of the finite element method [196]. Here the material is divided into a number of subdomains, termed elements, with associated nodes. The elements are considered to consist of materials, the constitutive equations of which are well known, and, upon change of the system, the nodes suffer nodal displacements and concomitant generalized nodal forces. The method involves construction of a global stiffness matrix that comprises the contributions from all elements, the relevant boundary conditions and body and thermal forces; a typical problem is then to compute the nodal displacements (i. e., the local strains) by solving the system $\mathbf{K} \cdot \mathbf{u} = \mathbf{F}$, where $\mathbf{K}$ is the stiffness matrix, $\mathbf{u}$ the vector of nodal displacements, and **F** the vector of generalized nodal forces, for **u**, most often by inverting **K**. This equation can be derived from the principle of minimum energy. The method, which has been discussed in a large number of excellent books and reviews, is elegant and powerful. However, it fails when the behavior of heterogeneous solids is targeted, e. g., that of composites comprising a large number of inclusions in a continuous matrix.

An alternative approach to the finite element approach is one, introduced as a concept by Courant as early as 1943 [197], in which the total energy functional, implicit in the finite element method, is directly minimized with respect to all nodal positions. The approach is conjugate to the finite element method and merely differs in its procedural approach. It parallels, however, methods often used in atomistic modeling schemes where the potential energy functional of a system (e. g., given by the "force field") is minimized with respect to the position of all (or at least many) atoms of the system. A simple example of this emerging technique is given below.

7.2 The Representative Volume Element Size in Elastic Composites: A Numerical Study

Locally, composites are heterogeneous. However, sufficiently large laboratory samples behave homogeneously. The problem is, therefore, to decide whether a given finite element is representative or not for describing the overall constitutive behavior of the composite studied. This question is a central one for a micromechanics-based non-local constitutive description [198] of random composites. It is also crucial for numerical prediction [199] of the overall constitutive behavior based on a known microstructure and constituent properties [199]. To date, the question studied using this "new" technique is the approach to the uniform representative volume element (RVE) size in glass/epoxy elastic composites of a random dispersion of non-overlapping identical spheres [199] and parallel identical cylinders [200]. A home-built, displacement-based, constant-strain tetrahedral element code [199] exists which was employed to numerically predict the overall elastic moduli and underlying local stresses in numerous Monte Carlo (MC) realizations of these two elastic composites. About a hundred MC realizations (see Fig. 7.1) with up to 64 non-overlapping identical spheres [199] and about 1000 with up to 289 perfectly aligned fibers [200] were considered. Perfect adhesion between matrix and inclusions was assumed and three-dimensional periodic meshes with up to several million elements were employed. The overall strain was imposed by changing the unit cell's continuation vectors. A conjugate-gradient iterative solver was used to minimize the total energy of the deformed models [199].

For the composite of a random dispersion of non-overlapping identical spheres, the average overall elastic constants calculated with only 8 spheres in the unit cell were already practically indistinguishable from those calculated with 64 spheres (see Fig. 7.2). Moreover, the scatter of the individual

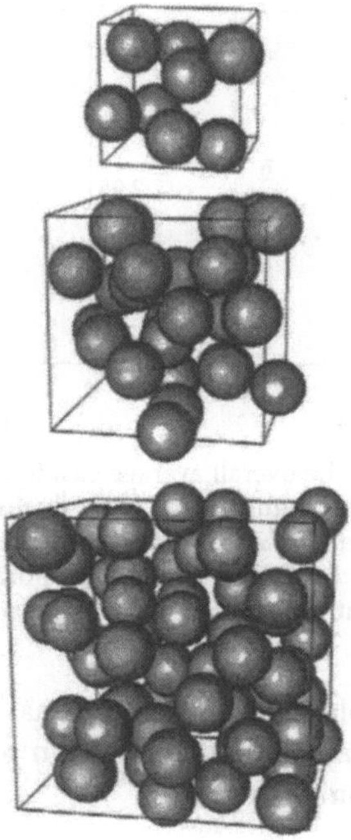

Fig. 7.1. Sketches of periodic MC realizations with 8, 27, and 64 spheres in disordered unit cells. Seven MC realizations with 8 spheres per unit cell, 4 realizations with 27, and 1 with 64 were employed. All microstructures studied had the same inclusion volume fraction of 0.27

estimates was very small. Thus, based on only a few MC realizations, one can accurately predict the overall elastic constants of this random composite.

The homogenization was also found [200] to be incredibly fast for both longitudinal and transverse overall elastic moduli of the transversely-random [201] unidirectional composite of parallel fibers. It was almost impossible to distinguish [200] the overall ensemble-average moduli calculated with 9 and 289 fibers in a disordered unit cell.

Assuming a matrix-dominated failure and based on the von Mises criterion for the epoxy matrix, the overall transverse failure strain of the MC realization of the unidirectional glass/epoxy composite was predicted [200]. Here, based on the nodal displacements available at the end of the conjugate-gradient minimization runs, the local stresses induced in the matrix by an overall transverse straining of the MC realization were studied. Drastic differences between the predictions obtained with the disordered MC realizations and those obtained with the traditional regular square or hexagonal models [201] were found. Quantitatively, the square and hexagonal packing arrays predict that the overall transverse strain to failure is about 1.1%, while the disordered MC realizations with 289 fibers in the unit cell already failed between 0.39 and 0.43%, in excellent agreement with common experimental practice. It appears,

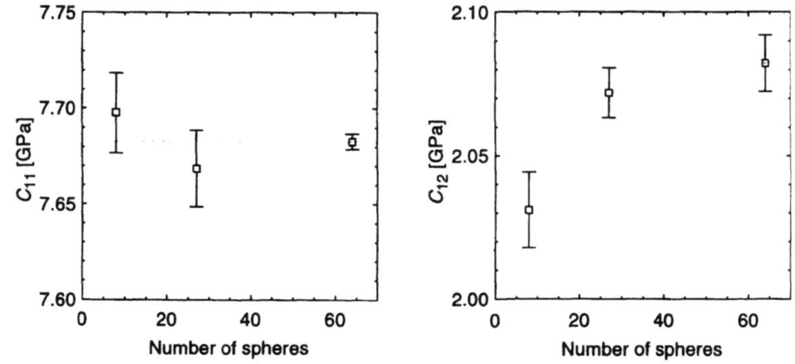

Fig. 7.2. Numerical estimates of the overall average elastic constants. The error-bars were calculated as $\sigma \cdot N^{0.5}$, where N is the number of individual estimates for each of the two overall elastic constants obtained with a given number of spheres in the unit cell and σ is the standard deviation. The two horizontal dashed lines drawn through the averages obtained with 64 spheres are meant to facilitate the convergence analysis

therefore, that it is impossible to use the traditional regular square or hexagonal models to predict the transverse failure strain of actual, transversely-random [201] unidirectional composites.

8 Conclusions and Outlook

In this review, the state of the art of the bridging of the gap between quantum chemical, atomistic, coarse-grained (and almost macroscopic) models of polymers has been discussed. Simulations with coarse-grained models provide the promise of the equilibration of models of dense amorphous polymers, whereas such equilibration is extremely difficult if the models are expressed in fully atomistic detail. The review presents the status of this rapidly developing field as of the beginning of 1998. A few minor additions were incorporated in the page proof, early in 2000, in response to suggestions from the reviewer.

The creation of suitable coarse-grained representations of a fully atomistic model requires compromises, which can be made in different ways. Useful coarse-grained models produced with different compromises are described here. Some of these approaches have been followed for more than six years – e.g. the attempt (Sect. 5) of a “mapping”, where input from chemically realistic models is used for effective potentials of the bond fluctuation model in order to “mimic” real polymers by a lattice model and, more recently, in an alternative approach in continuous space (Sect. 6) – while others are very recent, such as the novel methods to generate configurations of dense polymer systems (Sect. 3) or the inverse mapping procecures (Sects. 4.4 and 6.3). In addition, procedures are emerging, which extrapolate even further into the (semi-)macroscopic regime (Sects. 6.4 and 7). This review article can hence only provide a kind of “snapshot picture” of the current state of a field where much work is still in progress, and important further developments for many of the methods desribed here are anticipated. However, despite all the "missing links", we nevertheless feel that it is very useful to give a summary of the present status, which allows a clearer identification of the most promising routes that should be pursued in the future. This is the general spirit that has been followed here.

In the following, we give some more specific comments on the individual sections of this review. Section 2 is devoted to the smallest scales, discussing the problems of using effective force fields from quantum chemistry in classical simulations, and the possibilities for alternative routes, such as the Car-Parrinello approach, which is closer to a first-principles approach than the use of heuristic pair potentials, bond-angle potentials, etc., but which also suffers from its own limitations. In making the connection between coarse-grained models at various levels of coarse-graining and the real world, some atomistic input is inevitably needed, and hence the rapid progress described in this section will have a broad impact.

The feasibility of a vector field method has been demonstrated. Once the correct vector field has been generated, several configurations of any desired density of linear chains can be generated with ease. Another robust and efficient methodology for generating a very "reasonable" initial guess structure for atomistic simulations is the embedding algorithm conjugated with the novel ParRot move.

All these methods are rather new and not yet fully exploited, but we foresee widespread applications: the vector field method is an alternative starting point for generating states for a coarse-grained description of dense polymer systems, where more chemical detail can be inserted by inverse mapping procedures, for example. However, the resulting state of *any* inverse mapping procedure needs some further equilibration to allow relaxation of locally unfavorable configurations. The ParRot algorithm seems to be a very promising proposition to complement the existing machinery of Monte Carlo moves, and some of the drawbacks associated with more standard moves could possibly be avoided.

Based on the RIS Ansatz, the embedding algorithm benefits from a great flexibility in the choice of the input parameters that account for the local chain energy configuration; the input for the correlations of torsion angles along the chain backbones can be either calculated with the help of a force field, or extracted from measurements, or even biased in order to study any thinkable structural properties of the macromolecules.

Coarse-grained models can be generated on high-coordination lattices which are sparsely occupied even when the models represent an amorphous polymer system at real densities. The amount of structural detail that is sacrificed in the coarse graining (or the number of beads retained for representation of a specific real atomistic chain) can be varied using different methods for mapping onto the lattice, and different choices for the high coordination lattices. For polyethylene, one bead might be retained on the lattice for every 2–5 CH_2 groups. Coarse graining can also be accomplished in continuous space, without recourse to a lattice. None of the coarse-grained chains is expressed in atomistic detail, yet it is possible to retain distinctions between pairs of polymers, such as bisphenol-A-polycarbonate vs. polyethylene, or isotactic polypropylene vs. syndiotactic polypropylene, and to retain the distinctive features of different modifications of polycarbonate.

Useful results are often accessible from the coarse-grained representations themselves. Examples are provided by properties such as the cohesive energy density and surface energies. However, when greater structural fidelity is required, recourse can be made to "reverse-mapping", during which all of the missing atoms are restored to the coarse-grained model. This step would be required if, for example, there were a need for local intramolecular detail (such as the probability for occupancy of different values of the torsion angle at the real bonds in the chain) or local intermolecular detail (as in pair correlation functions for the atoms in the real chain).

A rather crude, but nevertheless efficient and successful, approach is the bond fluctuation model with potentials constructed from atomistic input (Sect. 5). Despite the lattice structure, it has been demonstrated that a rather reasonable description of many static and dynamic properties of dense polymer melts (polyethylene, polycarbonate) can be obtained. If the effective potentials are known, the implementation of the simulation method is rather straightforward, and also the simulation "data analysis" presents no particular problems. Indeed, a wealth of results has already been obtained, as briefly reviewed in this section. However, even this conceptually rather simple approach of coarse-graining (which historically was also the first to be tried out among the methods described in this article) suffers from severe bottlenecks – the construction of the effective potential is neither unique nor easy, and still suffers from the important defect that it lacks an intermolecular part, thus allowing only simulations at a given constant density.

Some limitations just described are avoided by an elegant generalization of the mapping procedure from the lattice to the continuum (Sect. 6). The key idea is to associate the logarithm of the probability distributions of suitable degrees of freedom of the input atomistic model directly with the effective potentials. Also, the off-lattice coarse-grained model allows the re-introduction of atomistic detail by inverse mapping in a fairly natural way. The coarse-grained model without this inverse mapping already allows very promising applications, such as a correlation between the Vogel-Fulcher parameters of various polycarbonate modifications and their chemical structure. Using inverse mapping, the static structure (as known from neutron scattering studies of fully and partially deuterated samples) can be also reproduced rather accurately, and thus the well-known standard "amorphous cell" method finds a powerful alternative.

As has already been emphasized in Fig. 1.1, there is the further problem of connecting the mesoscopic scale, where one considers length scales from the size of effective monomers to the scale of the whole coils, to still much larger scales, to describe structures formed by multichain heterophase systems. Examples of such problems are polymer blends, where droplets of the minority phase exist on the background of the majority matrix, etc. The treatment of these scales is really just in its infancy but, nevertheless, two very promising directions are briefly mentioned here: one is the idea of representing whole coils as effective deformable ellipsoids (Sect. 6.4), the other is a new view on the mechanical properties of such heterogeneous polymeric solids (Sect. 7).

The variety of approaches presented here provides an investigator with several options in the construction of a simulation. The "best" approach may very well depend on the structure of the real polymer of interest, and perhaps also on the polymer properties that the goal of the simulation.

What is clear is that the overall strategy, in which the computationally intensive equilibration is performed with a model that does not have atomistic detail, but which can nevertheless be unambiguously identified with a specific real system with atomistic detail, markedly increases the utility of simulations for polymer science. It provides access to much larger systems, and to phenomena that occur on much longer time scales than could be investigated with simulations in which full atomistic detail is retained throughout.

One of the important limitations of these methods, in their current state of development, is the necessity to conduct a new coarse-graining for each new type of polymer that is simulated. Major opportunities to increase the utility of this class of simulations lie in the development of methods, which make the mapping to the coarse-grained structure (and the "reverse-mapping" back to the structure expressed in fully atomistic detail) as general (and painless) as possible when new polymers are treated. This area is the subject of continuing work by all four groups.

Acknowledgments. This manuscript is the product of a series of meetings of the four research groups, or subsets thereof, over a period of three years. The authors thank both the agencies that funded the research, and also the agencies that supported the travel expenses necessary for this specific international collaboration. The investigators at ETH Zürich are indebted to the Swiss National Science Foundation (Nationalfonds) for financial support and to the Swiss Centre for Scientific Computing (SCSC) for significant amounts of computational resources. The US investigators gratefully acknowledge support from the National Science Foundation (DMR 9220369, DMR 9523278 and INT 9421585). The group at Universität Mainz acknowledges partial support from the Deutsche Forschungsgemeinschaft (DFG), grant No. Bi 314/12, and from the Bundesministerium für Bildung, Wissenschaft, Forschung and Technologie (BMBF), grants No. 03M4028, 03M40708, and BAYER AG. Work by the group at the Max-Planck-Institut was supported by the BMBF under grant No. 03 M 4070 8 and the Bayer Corp. The project has benefited from useful interactions on various aspects of the work with J. Cho, B. M. Forrest, W. F. van Gunsteren, T. Haliloğlu, D. W. Heermann, M. Müller, R. F. Rapold, D. Richter and his group, F. Schmid, W. Tschöp, J. Wittmer, M. Wolfgardt, M. M. Zehnder, and the Bayer polymer physics group (especially J. Batoulis). For bringing this manuscript in final form we are grateful to Prof. G. Allegra and his coworkers for numerous comments.

References

1. Cohen MH, Turnbull DJ (1959) J Chem Phys 31: 1164
2. Fujita H (1961) Fortschr Hochpolym Forsch 3: 1
3. Stern SA, Fang SM, Frisch HL (1972) Polym Sci A2 10: 201
4. Vrentas JS (1977) J Polym Phys Ed 15: 403

5. Vrentas JS, Duda JL (1978) J Appl Poly Sci 22: 2325
6. Shah VM, Stern SA, Ludovice PL (1989) Macromolecules 22: 4660
7. Binder K (ed) (1979) Monte Carlo methods in statistical physics, Springer, Berlin
8. Ciccotti G, Hoover WG (eds) (1986) Molecular dynamics simulations of statistical mechanical systems. North-Holland, Amsterdam
9. Allen MP, Tildesley DJ (1987) Computer simulation of liquids. Clarendon Press, Oxford
10. Binder K (ed) (1992) Monte Carlo methods in condensed matter physics. Springer, Berlin Heidelberg New York
11. Allen MP, Tildesley DJ (ed) (1993) Computer simulation in chemical physics. Kluwer, Dordrecht
12. Binder K (ed) (1995) Monte Carlo and molecular dynamics simulations in polymer science. Oxford University Press, Oxford
13. Binder K, Ciccotti G (eds) (1996) Monte Carlo and molecular dynamics of condensed matter. Società Italiana di Fisica, Bologna
14. Frenkel D, Smit B (1996) Understanding molecular simulation: from algorithms to applications. Academic Press, New York
15. Kremer, K In: Binder K, Ciccotti G (eds) (1996) Monte Carlo and molecular dynamics of condensed matter. Società Italiana di Fisica, Bologna
16. Kremer K, Binder K (1988) Comput Phys Repts 7: 259
17. Binder K (1991) Macromol Chem: Macromol Symp 50: 1
18. Binder K (1992) In: Bicerano J (ed) Computational modeling of polymers. Marcel Dekker, New York, p 221
19. Kremer K (1993) In: Allen MP, Tildesley DJ (ed) (1993) Computer simulation in chemical physics. Kluwer, Dordrecht
20. Binder K, Paul W (1997) J Polym Sci B 35: 1
21. Rouse PE (1953) J Chem Phys 21: 1272
22. Doi M, Edwards S (1986) Theory of polymer dynamics. Clarendon Press, Oxford
23. de Gennes PG (1971) J Chem Phys 69: 572
24. Ferry, JD (1980) Viscoelastic properties of polymers. John Wiley, New York
25. McKenna, GB (1990) In: Booth C, Price C (eds) Comprehensive polymer science, vol 2, Polymer Properties. Pergamon, Oxford, p 311
26. Roe RJ (1991) Computer simulation of polymers. Prentice Hall, NY
27. Baschnagel J, Binder K, Paul W, Laso M, Suter U, Batoulis I, Jilge W, Bürger T (1991) J Chem Phys 95: 6014
28. Paul W, Binder K, Kremer K, Heermann DW (1991) Macromolecules 24: 6332
29. Baschnagel J, Qin K, Paul W, Binder K (1992) Macromolecules 25: 3117
30. Paul W, Pistoor N (1994) Macromolecules 27: 1249
31. Zimmer KM, Linke A, Heermann DW, Batoulis J, Bürger T (1996) Macromol Theory Simul 5: 1065
32. Tries V, Paul W, Baschnagel J, Binder K (1997) J Chem Phys 106: 738
33. Rutledge GC, Suter UW (1991) Macromolecules 24: 1921
34. Rutledge GC, Suter UW, Papaspyrides (1991) Macromolecules 24: 1934
35. Flory PJ (1989) Statistical mechanics of chain molecules. Hanser, New York
36. Mattice WL, Suter UW (1994) Conformational theory of large molecules. John Wiley, New York
37. Carmesin I, Kremer K (1988) Macromolecules 21: 2819
38. Wittmann HP, Kremer K (1990) Comp Phys Comm 61: 309, (1992) Comp Phys Comm 71: 343
39. Deutsch HP, Binder K (1991) J Chem Phys 94: 2294
40. Paul W, Binder K, Heermann DW, Kremer K (1991) J Phys II (France) 1: 37
41. Paul W, Binder K, Heermann DW, Kremer K (1991) J Chem Phys 95: 7726
42. Wittmer J, Paul W, Binder K (1994) J Phys II (France) 4: 873

43. Tschöp W, Kremer K, Batoulis J, Bürger T, Hahn O (1998) Acta Polymerica 49: 61
44. Tschöp W, Kremer K, Batoulis J, Bürger T, Hahn O (1998) Acta Polymerica 49: 75
45. Murat M, Kremer K, J Chem Phys (1998) 108, 4340
46. Müller-Plathe F (1997). In: Wilson S, Diercksen GHF (eds.) Problem solving in computational molecular science: molecules in different environments, NATO ASI Series B Physics. Kluwer, Dordrecht
47. Pèrez-Jordà JM, Becke AD (1995) Chem Phys Lett 233: 134
48. Jørgensen P, Oddershede J, Beebe NHF (1978) J Chem Phys 68: 2527
49. van Gisbergen SJA, Snijders JG, Baerends EJ (1995) J Chem Phys 103: 9347
50. Stouch TR, Williams DE (1993) J Comp Chem 14: 858
50A. Koch U, Stone AJ (1996) J Chem Soc Faraday Trans 92: 1701
51. Müller-Plathe F (1996) Mol Sim 18: 133
52. Liu H, Müller-Plathe F, van Gunsteren WF (1995) J Am Chem Soc 117: 4363
53. Hünenberger PH, van Gunsteren WF (1997). In: van Gunsteren WF, Weiner PK, Wilkinson AJ (eds.), Computer simulation of biomolecular systems - theoretical and experimental applications, vol 3. Escom Science Publishers, Leiden, 1997
54. Njo SL, van Gunsteren WF, Müller-Plathe F (1995), J Chem Phys 102: 6199
55. Berweger CD, van Gunsteren WF, Müller-Plathe F (1995) Chem Phys Lett 232: 429
56. Ulrich P, Scott WRP, van Gunsteren WF, Torda AE (1997) Proteins 27: 367
56A. Faller R, Schmitz H, Biermann O, Müller-Plathe F (1999) J Comp Chem 20: 1009
57. Müller-Plathe F (1994) Braz J Phys 24: 965
58. Nezel T, Müller-Plathe F, Müller MD, Buser HR (1997) Chemosphere 35: 1895
59. Röthlisberger U, Laasonen K, Klein ML (1996) J Chem Phys 104: 3692; Sprik M, Röthlisberger U, Klein ML (1997) J Phys Chem B 101: 2745
60. Müller-Plathe F (1994) Acta Polym 45: 259
61. Neyertz S, Brown D, Thomas JD (1994) J Chem Phys 101: 10054
62. Smith GD, Yoon DY, Jaffe RL, Colby RH, Krishnamoorti R, Fetters LJ (1996) Macromolecules 29: 3462
63. Müller-Plathe F (1997) Polymer 38: 2259
64. Müller-Plathe F, van Gunsteren WF (1994) Macromolecules 27: 6040
65. Müller-Plathe F, Liu H, van Gunsteren WF (1995) Comput Polym Sci 5: 89
65A. Warshel A, Levitt M (1976) J Mol Biol 103: 227
65B. Field MJ, Bash PA, Karplus M (1990) J Comp Chem 6: 700
66. Bakowies D, Thiel W (1996) J Phys Chem 100: 10580; Bakowies D, Thiel W (1996) J Comp Chem 17: 87
67. Müller-Plathe F (1997) Adv Quant Chem 28: 82
68. Liu H, Müller-Plathe F, van Gunsteren WF (1996) Chem Eur J 2: 191
69. Liu H, Müller-Plathe F, van Gunsteren WF (1996) J Mol Biol 261: 454
70. Berweger CD, van Gunsteren WF, Müller-Plathe F (1997), J Comp Chem 18: 1484
71. Berweger CD, van Gunsteren WF, Müller-Plathe F, J (1998) Chem Phys, 108: 8773; Berweger CD, van Gunsteren WF, Müller-Plathe F (1999) Angew Chem Int Ed 38: 2609; Berweger CD, van Gunsteren WF, Müller-Plathe F (1999) J Chem Phys 111: 8987
72. Waldeck DH (1991) Chem Rev 91: 415
73. Chen W, Hase WL, Schlegel HB (1994) Chem Phys Lett 228: 436
74. Hartke B, Carter EA (1992) J Chem Phys 97: 6569
75. Tuckerman M, Laasonen K, Sprik M, Parrinello M (1995) J Chem Phys 103: 150
76. Nusterer E, Blöchl PE, Schwarz K (1996) Angew Chem 108: 187
77. Röthlisberger U, Sprik M, Klein M (1998) J Chem Soc Faraday Trans 94: 501
78. Montanari B, Jones RO (1997) Chem Phys Lett 272: 347
79. Meirovitch HJ (1982) J Phys A 15: L735
80. Theodorou DN, Suter UW (1985) Macromolecules 18: 1206
81. Theodorou DN, Suter UW (1985) Macromolecules 18: 1467

82. McKechnie JI, Brown D, Clarke JH (1992) Macromolecules 25: 1562
83. Zimmer KM, Heermann DW (1995) J Comput-Aided Mater Des (1995) 2: 1
84. Gusev AA, Zehnder MM, Suter UW (1994) Macromolecules 27: 615
85. Kotelyanskii M, Wagner NJ, Paulaitis ME (1996) Macromolecules 29: 8497
86. Gusev AA, Suter UW (1991) Phys Rev A 43: 6488
87. Müller-Plathe F (1991) Macromolecules 24: 6475
88. Sok RM, Berendsen HJ, van Gunsteren WF (1992) J Chem Phys 96: 4699
89. Müller-Plathe F, Rogers SC, van Gunsteren WF (1992) Chem Phys Lett 199: 237
90. Pant PV, Boyd RH (1993) Macromolecules 26: 679
91. Gusev AA, Arizzi S, Suter UW, Moll DJ (1993) J Chem Phys 99: 2221
92. Gusev AA, Suter UW (1993) J Chem Phys 99: 2228
93. Gusev AA, Suter UW, Moll DJ (1995) Macromolecules 28: 2582
94. Knopp B, Suter UW, Gusev AA (1997) Macromolecules 30: 6107
95. Knopp B, Suter UW (1997) Macromolecules 30: 6114
96. de Gennes PG (1979) Scaling concepts in polymer physics. Cornell University, Ithaca
97. Kleinert H (1990) Path integrals in quantum mechanics, statistics and polymer physics. World Scientific, Singapore
98. Santos S (1998) Dissertation ETH Nr. 12941; Santos S, Suter UW, in preparation
99. Porod G (1949) Monatsh. Chem 80: 251
100. Lifschitz EM, Landau LD (1986) Course of theoretical physics, vol 5. Pergamon, London
101. Edwards SF (1970) Discuss. Faraday Soc. 49: 43
102. Freed KF (1971) J Chem Phys 54: 1453
103. Bawendi MG, Freed KF (1985) J Chem Phys 83: 2491
104. Mortenson ME (1985) Geometric modelling, John Wiley, New York
105. Metropolis N, Rosenbluth AW, Rosenbluth AH, Teller AH, Teller E (1953) J Chem Phys 21: 1087
106. Kirkpatrick S, Gelat CD, Vecchi MP (1983) Science 220: 671
107. Binder K (1987) Applications of the Monte Carlo method in statistical physics. In: Topics in Current Physics, Springer, Berlin
108. Binder K, Heermann DW (1988) Monte Carlo simulation in statistical physics. In: Series in Solid-State Sciences, Springer Berlin
109. Aarts E, Korst J (1989) Simulated annealing and Boltzmann machines, John Wiley, England
110. Geman S, Geman D (1984) IEEE Trans Patt Anal Mach Intell 6: 721
111. Ingber L (1989) Math Comp Modelling 12: 967
112. Ingber L, Rosen B (1992) Math Comp Modelling 16: 87
113. Parks GT (1990) Nuclear Technology 89: 233
114. Care CM (1996) J Phys A: Math Gen 29: L505
115. Müller M, Santos S, Nievergelt J, Suter UW, J Chem Phys, in print
116. Criven GM, Havel TF (1988) Distance geometry and molecular conformation, Research Studies Press Ltd, Taunton
117. Wako H, Scheraga HA (1982) J Protein Chem 1: 85
118. Havel TF, Wuthrich K (1984) Bull Math Biol 46: 281
119. Billeter M, Havel TF, Wuthrich K (1986) J Comp Chem 8: 132
120. Go N, Scheraga HA (1970) Macromolecules 3: 178
121. Flory PJ (1974) Macromolecules 7: 381
122. Kitao A, Go NJ (1991) J Comput Chem 12: 359
123. Gray CG, Gubbins KE (1984) Theory of molecular fluids, vol 1. Clarendon Press, Oxford
124. Rehahn M, Mattice WL, Suter UW (1997) Rotational isomeric state models in macromolecular systems. In: Advances in Polymer Science, Springer, Berlin

125. Tomaselli M, Robyr P, Meier BH, Grob-Pisano C, Ernst RR, Suter UW (1996) Mol Phys 89: 1663
126. Tomaselli M, Zehnder MM, Robyr P, Grob-Pisano C, Ernst RR, Suter UW (1997) Macromolecules 30: 3579
127. Santos S, Müller M, Nievergelt J, Suter UW, J Chem Phys, in print
128. Lal M (1969) Mol Phys 17: 57
129. Bishop M, Ceperley D, Frisch HL, Kalos MH (1982) J Chem Phys 76: 1557
130. Vacatello M, Avitabile G, Corradini P, Tuzi A (1979) Conv Ital Sci Macromol 4: 135
131. Madras N, Sokal AD (1988) J Stat Phys 50: 109
132. Wittmann HP, Kremer K, Binder K (1988) J Chem Phys 96: 6291
133. Boyd RH (1989) Macromolecules 22: 2477
134. Siepman JI (1990) Mol Phys 70: 1145
135. Frenkel D, Mooij GCAM, Smit B (1991) J Phys Condens Matter 3: 3053
136. Siepman JI, Frenkel D (1992) Mol Phys 75: 59
137. Almarza NG, Enciso E, Bermejo FJ (1992) J Chem Phys 96: 4625
138. de Pablo JJ, Laso M, Suter UW (1992) J Chem Phys 96: 2395
139. Go N, Scheraga HA (1976) Macromolecules 9: 535
140. Dodd LR, Boone TD, Theodorou DN (1993) Mol Phys 78: 961
141. Leontidis E, de Pablo JJ, Laso M, Suter UW (1994) Atomistic modelling of physical properties. In: Advances in Polymer Science, Springer, Berlin
142. Ryckaert JP, Bellemans A (1975) Chem Phys Let 30: 123
143. Abe A, Jernigan RL, Flory PJ (1966) J Am Chem Soc 88: 631
144. Doruker P, Mattice WL (1997) Macromolecules 30: 5520
145. Smith GD, Yoon DY, Jaffe RL (1995) Macromolecules 28: 5397
146. Paul W, Yoon DY, Smith GD (1995) J Chem Phys 103: 1702
147. Skolnick J, Kolinski A (1991) J Mol Biol 221: 499
148. Kolinski A, Skolnick J (1992) J Chem Phys 97: 9412
149. Kolinski A, Godzik A, Skolnick J (1993) J Chem Phys 98: 7420
150. Godzik A, Kolinski A, Skolnick J (1993) J Comput Chem 14: 1194
151. Rapold RF, Mattice WL (1995) J Chem Soc Faraday Trans 91: 2435
152. Raghunathan G, Jernigan RL (1997) Protein Science 6: 2072
153. Rapold RF, Mattice WL (1996) Macromolecules 29: 2457
154. Doruker P, Rapold RF, Mattice WL (1996) J Chem Phys 104: 8742
155. Bahar I, Cho J, Doruker P, Erman B, Haliloglu T, Kim EG, Mattice WL, Monnerie L, Rapold RF (1997) Trends Polym Sci (Cambridge, UK) 5: 155
156. Haliloglu T, Mattice WL (1998) J Chem Phys 108: 6989
157. Suter UW, Pucci S, Pino P (1975) J Am Chem Soc. 97: 1018
158. Haliloglu T, Cho J, Mattice WL (1998) Macromol Theory Simul 7: 613
158A. Haliloglu T, Mattice WL (1999) J Chem Phys 111: 4327
159. Post CB, Zimm BH (1979) Biopolymers 18: 1487
160. Cho J, Mattice WL (1997) Macromolecules 30: 637
161. Hirschfelder JO, Curtiss CB, Bird RB (1954) Molecular theory of gases and liquids. John Wiley, New York
162. Doruker P, Mattice WL (1998) Macromolecules 31: 1418
163. van Krevlan DW (1990) Properties of polymers. Elsevier Science, Amsterdam, The Netherlands
164. Bicerano J (1993) Prediction of polymer properties. Marcel Dekker, Inc, p 117
165. Smith GD, Yoon DY, Zhu W, Ediger ME (1994) Macromolecules 27: 5563
166. Honnell KG, McCoy JD, Curro JG, Schweizer KS, Narten AH, Habenschuss A (1991) J Chem Phys 94: 4659
167. Doruker P, Mattice WL (1998) Macromol Symp 133: 47
168. Paul W, Smith GD, Yoon DY (1997) Macromolecules 30: 7772
169. von Meerwall ED, Beckman S, Jang JH, Mattice WL (1998) J Chem Phys 108: 4299

170. Ozisik R, Mattice WL, work in progress
171. Mattice WL (1976) Macromolecules 8: 48
172. Misra S, Fleming PD, Mattice WL (1995) J Comput-Aided Mater Des 2: 101
173. Jang JH, Mattice WL (1998) Polym Commun 40: 1911
174. Wilkes CE, Folt VL, Krimm S (1973) Macromolecules 6: 235
175. Paul W (1992) AIP Conf Proc 256: 145
176. Paul W, Binder K, Batoulis I, Pittel B, Sommer KH (1993) Macromol Chem, Macromol Symp 65: 1
177. Kremer K, Grest GS (1990) J Chem Phys 92: 5057
178. Pearson DS, Ver Strate G, von Meerwall E, Schilling FC (1987) Macromolecules 20: 1133
179. des Cloizeaux J (1993) J Phys I (Paris) 3: 1523
180. de Gennes PG (1967) Physics (NY) 3: 97
181. Richter D, Butera R, Fetters LJ, Huang JS, Farago B, Ewen B (1992) Macromolecules 25: 6156
182. Rigby DJ, Roe RJ (1987) J Chem Phys 87: 7285
183. Smith GD, Yoon DY (1994) J Chem Phys 100: 649
184. Tries V (1996) Dissertation, Johannes-Gutenberg Universität Mainz, unpublished, Baschnagel J, Paul W, Tries V, Binder K (1998) Macromolecules 31: 3856
185. Eilhard J (1996) Untersuchungen zur Struktur und Dynamik von Polykondensaten, Jül-Bericht, Jülich; Eilhard J, Zirkel A, Tschöp W, Hahn O, Kremer K, Scharpf O, Richter D, Buchenau U (1999) J Chem Phys 110: 1819
186. Final report of the BMFT project No 03M4038 (1992) "Theoretisch-experimentelle Struktur-Eigenschaftskorrelation von Polykondensaten" (unpublished); Sommer K, Batoulis J, Jilge W, Morbitzer L, Pittel B, Plaetschke R, Reuter K, Timmermann R, Binder K, Paul W, Gentile FT, Heumann DW, Kremer K, Laso M, Suter UW, Ludovice PJ (1991) Adv Materials 3: 590
187. Vogel H (1921) Phys Z 22: 642; Fulcher GS (1925) J Am Ceram Soc 8: 339
188. Wall FT, Mandel F (1975) J Chem Phys 63: 4592
189. Wolfgardt M, Baschnagel J, Binder K (1995) J Phys II (France) 5: 1035
190. Pearson DS, Fetters LJ, Graessley WW, ver Strate G, von Meerwall E (1994) Macromolecules 27: 711
191. Wolfgardt M, Binder K (1996) Macromol Theory Simul 5: 699
192. Verlet L (1967) Phys Rev 165: 98
193. Sears VF (1984) Thermal-neutron scattering lengths and cross-sections for condensed matter research. Chalk River Nuclear Laboratories, Chalk River
194. Suter UW cited in Ref. 183
195. Tschöp W, Ph. D Thesis, Univ of Bonn (1998)
196. Turner MJ, Clough RW, Martin HC, Topp LJ (1956) J Aeron Sci 23: 805
197. Courant R (1943) Bull Am Math Soc 49: 1
198. Drugan WJ, Willies JR (1996) J Mech Phys Solids 44: 497
199. Gusev AA (1997) J Mech Phys Solids 45: 1449
200. Gusev AA, Hine PJ, Ward IM, Composite Sci Tech, in print
201. Adams DF, Tsai SW (1969) J Compos Mater 3: 368

Received: February 1998

Principles of the Quantitative Description of the Chemical Structure of Synthetic Polymers

Semion I. Kuchanov

Keldysh Institute of Applied Mathematics, Russian Academy of Sciences,
Miusskaya Square 4, 125047 Moscow, Russia
e-mail: kuchanov@orc.ru

A brief introduction to the statistical chemistry of polymers is provided and the potential applications of this approach for the mathematical modeling of some important processes of the synthesis of high molecular weight compounds are demonstrated. Proceeding from the consideration of the main peculiarities of synthetic polymers, the statistical characteristics of their chemical structure generally used for its quantitative description are indicated. The principle kinetic models of macromolecular reactions currently employed in polymer chemistry are considered. The advantages and disadvantages of different methods applied for the calculation of the composition and chemical structure of macromolecules are discussed. The results of invoking these methods in order to provide a quantitative description of the products of linear chain copolymerization and branched polycondensation are reported.

Keywords. Statistical chemistry of polymers, Chain and step polymerizations, Linear and branched polymers

List of Symbols and Abbreviations

m number of types of monomeric units
l_α number of α-type units in a macromolecule
$\mathrm{M}_\alpha, M_\alpha$ symbol of α-type monomer and its concentration
$\overline{\mathrm{M}}_\alpha$ symbol of α-type monomeric unit
l degree of polymerization of a macromolecule
$\nu_{\alpha\beta}$ element of probability transition matrix
ζ composition vector
$f(l)$ size distribution (SD)
$f(\mathbf{l})$ size composition distribution (SCD)
X_α average mole fraction of α-type units in a copolymer
$\lambda_{\alpha\beta}$ element of covariance matrix of SCD
ΔH enthalpy of mixing of a copolymer
T absolute temperature
$\chi_{\alpha\beta}$ the Flory parameter of pair interaction between units $\overline{\mathrm{M}}_\alpha$ and $\overline{\mathrm{M}}_\beta$
U_k sequence involving k monomeric units
$P\{U_k\}$ mole fraction of sequence k
T_g glass transition temperature
K_M microgeterogeneity coefficient
$Y_{\alpha\beta}(k)$ two-point chemical correlator
ω_s weight fraction of sol
ω_g weight fraction of gel
$\mathrm{R}_\alpha, R_\alpha$ symbol of α-type radical or ion and its concentration
$k_{\alpha\beta}$ constant of propagation reaction between R_α and M_β
$k^{\mathrm{t}}_{\alpha\beta}$ constant of termination reaction between R_α and R_β
$r_{\alpha\beta}, r_{\beta\alpha}$ reactivity ratios for binary free-radical copolymerization of monomers M_α and M_β
x_α mole fraction of monomer M_α

p	overall mole conversion of monomers
S_α	state of Markov chain
$\mathbf{Q}$	probability transition matrix
π	stationary vector of Markov chain
A_i	symbol of functional group whose type is i
Q_{ij}	symbol of chemical bond appearing due to reaction between groups A_i and A_j
z_{ij}	symbol of low molecular weight by-product formed for reaction between A_i and A_j
W	rate of condensation reaction
W_d	rate of degradation reaction
W_{out}	rate of removing by-product from reaction system
a_i	number of groups A_i in molecule
M	overall concentration of monomeric units
t	time
$C(\mathbf{l};\mathbf{a};t)$	size composition functionality distribution (SCFD)
μ_i	average number of groups A_i per one monomeric unit
$\mu^0{}_i$	initial value of μ_i
$f_W(\mathbf{l})$	weight SCD
$G_W(\mathbf{s})$	generating function (*gf*) of this distribution
ν_α	initial mole fraction of monomer M_α
$\mathbf{p}$	matrix of bonds
κ	stoichiometric matrix
$\mathbf{f}$	matrix of functionalities
P_W	weight average degree of polymerization
pi	conversion of groups A_i
FSSE	first-shell substitution effect
k_{ij}	rate constant of condensation reaction between A_i and A_j

1 Introduction – Specific Features of the Characterization of Polymers

Polymer products synthesized in laboratories and in industry represent a set of individual chemical compounds whose number is practically infinite. Macromolecules of such products can differ in their degree of polymerization, tacticity, number of branchings and the lengths that connect their polymer chains, as well as in other characteristics which describe the configuration of the macromolecule. In the case of copolymers their macromolecules are known to also vary in composition and the character of the alternation of monomeric units of different types. As a rule, it is impossible to provide an exhaustive quantitative description of such a polymer system, i.e. to indicate concentrations of all individual compounds with a particular chemical (primary) structure. However, for many practical purposes it is often enough to define a polymer specimen only in terms of partial distributions of molecules for some of their main characteristics (such as, for instance, molecular weight or composition) avoiding completely a description of their chemical structure. In other cases the latter plays a central role in predetermining some of the physicochemical properties, where a knowledge of the quantitative characteristics of the polymer microstructure is imperative.

The more detailed quantitative description of a polymer system can be achieved by extending the set of parameters by whose values the individual macromolecules are distinguished, the wider is the range of macroscopic properties of the system that it is possible to predict. The last circumstance is associated with the fact that the primary structure of macromolecules carries specific information on their possible conformations, secondary and supermolecular structures and, consequently, on their physical properties. The task of revealing quantitative correlations between these properties and the characteristics of a macromolecule primary structure which fall within the realms of statistical physics of polymers constitutes one of the three challenging problems whose solution is indispensable for scientifically grounded prediction in obtaining polymers with a desired set of properties. The second task, relevant to chemical kinetics of macromolecular reactions, consists in establishing the mechanism and measuring constants of elementary reactions of the processes of polymer synthesis. Once this problem has been solved, it becomes possible, in the framework of an adequately chosen kinetic model, to find quantitative dependencies of statistical characteristics of the primary polymer structure on the reactivities of the initial compounds and the synthesis conditions. Only this third problem, pertaining to the statistical chemistry of polymers, will be addressed in this review.

The simplest, from the viewpoint of topological structure, are the linear polymers. Depending on the number m of the types of monomeric units they differentiate homopolymers ($m=1$) and copolymers ($m \geq 2$). In the most trivial case molecules in a homopolymer are merely identified by the number l of monomeric units involved, whereas the composition of a copolymer macromolecule is defined by vector $\mathbf{l}$ with components $l_1,\ldots,l_\alpha,\ldots,l_m$ equal to the numbers of monomeric units of each type. At identical composition these molecules can vary in microstructure which is characterized by the manner of alternation of different units in a copolymer chain. Because the values of the average degree of polymerization $l=l_1+\ldots+l_m$ in synthetic copolymers normally constitute 10^2–10^4 it becomes clear that the number of conceivable types of isomers with different microstructure turns out to be practically infinite. Naturally, a quantitative description of any polymer specimen comprising macromolecules with such an impressive number of configurations can be performed exclusively by statistical methods.

In terms of a statistical approach to every macromolecule of a copolymer specimen there can be put in correspondence a certain realization of a stochastic process of conventional movement along a copolymer chain. This movement can be conveniently thought of as a sequence of random transitions from a unit of the chain to the neighboring one. Here the probability of finding a monomeric unit of a particular type at every step is predetermined by the stochastic process which describes the particular polymer specimen. To consider the set of trajectories corresponding to realizations of infinite length it is convenient to presume that upon reaching the terminal unit of a molecule the trajectory falls into the absorbing state to remain there from this moment on. Hence, for each particular specimen of a linear copolymer with m types of units, there is a corresponding stochastic process with a discrete time, which has an m transient and one absorbing state. A stochastic process of such a kind in discrete time with discrete state-space is referred to as a chain. The best known of these are the Markov chains [1] for which the probability $\nu_{\alpha\beta}$ to fall at any step into a state whose type is β is controlled solely by the type α of the state at the preceding step.

Only if the attempt to formulate a general algorithm to find the probability of any trajectory of the above stochastic process is a success, the statistical description of a copolymer specimen is considered to be exhaustive. For those copolymers where the alternation of units in the macromolecules is described by the Markov chain such an algorithm is trivial. Therefore, proceeding from the general theory it is possible to determine immediately any statistical characteristics of Markovian copolymers expressed through the elements $\nu_{\alpha\beta}$ of the probability transition matrix of the corresponding Markov chain. Essentially, any special features of particular processes of the synthesis of such copolymers are only taken into account when the dependence of matrix elements $\nu_{\alpha\beta}$ on time as well as on the stoichiometric and kinetic parameters of a reaction system have to be determined. Thus, the problem of mathematical modeling of the chemical structure of the products of the synthesis of copolymers which obey Markovian statistics proves to be quite trivial. That is why it is of prime importance for an engineer researcher to know already at the stage of choosing an adequate kinetic model of a particular process of obtaining a copolymer whether the latter will be Markovian. This question has already been answered for many cases of practical significance [2].

What kind of strategy should be adopted for the mathematical modeling of non-Markovian copolymers where the algorithm of finding the probability of any individual macromolecule is unknown? In this situation it is necessary to determine the statistical characteristics of the chemical structure of a polymer specimen for each specific method of its synthesis in the framework of the kinetic model chosen. These characteristics can be subdivided into two types. Those belonging to the first type describe the inhomogeneity of macromolecules for degree of polymerization l (chemical size) and/or composition, while the second type of characteristics gives an account of the arrangement of monomeric units in copolymer chains.

Characteristics of the first type include the weight size composition distribution (SCD) $f_W(\mathbf{l})$ which equals the probability for a unit chosen at random to belong to a macromolecule with given value of vector $\mathbf{l}$. Knowledge of this distribution is essential for the construction of the phase diagram of a solution or melt of a copolymer [3]; this diagram is of critical importance for predicting some of the properties of the copoymer, e.g. the transparency. The loss of the latter during the synthesis may occur due to the phase separation in the reaction system induced by chemical transformations. Such a phenomenon is caused by the rise in molecular inhomogeneity of a copolymer formed at high conversions. This assertion has rigorous theoretical substantiation [3, 4].

When dealing with high molecular weight polymers it is convenient instead of **l** to introduce the composition vector ζ with components $\zeta_\alpha = l_\alpha / l (\alpha = 1,\ldots,m)$ which, in combination with the chemical size l, is completely equivalent to **l**. For such copolymers the variables l and ζ_α may be thought of as continuous and recourse can be made to a expression which relates SCD $f_W(\mathbf{l})$ to the size distribution (SD) $f_W(l)$ and the composition distribution (CD) $W(\zeta \mid l)$ of macromolecules of given size l.

$$f_W(\mathbf{l}) = f_W(l) W(\zeta | l) \tag{1}$$

Applying a chromatographic method it is sometimes possible to separate copolymer molecules according to their size l and composition [5]. The SCD found in such a way can be compared with that calculated within the framework of the chosen kinetic model. The first- and second-order statistical moments of SCD are of special importance.

$$X_\alpha \equiv \bar{\zeta}_\alpha = \int_1^\infty f_W(l) dl \int_0^1 \cdots \int_0^1 \zeta_\alpha W(\zeta | l) d\zeta_1 \cdots d\zeta_m \tag{2}$$

$$\begin{aligned} \lambda_{\alpha\beta} &\equiv \overline{(\zeta_\alpha - X_\alpha)(\zeta_\beta - X_\beta)} \\ &= \int_1^\infty f_W(l) dl \int_0^1 \cdots \int_0^1 (\zeta_\alpha - X_\alpha)(\zeta_\beta - X_\beta) W(\zeta | l) d\zeta_1 \cdots d\zeta_m \end{aligned} \tag{3}$$

For example, the enthalpy of mixing of copolymer specimen ΔH per mole of monomeric units can be expressed

$$\frac{\Delta H}{RT} = -\sum_{\alpha<\beta}^{m} \chi_{\alpha\beta} \lambda_{\alpha\beta} \tag{4}$$

through the elements $\lambda_{\alpha\beta}$ (3) of the covariance matrix λ of its SCD and the Flory χ-parameters, values of which $\chi_{\alpha\beta}$ are available from the literature for many pairs of monomeric units $(\overline{\mathrm{M}}_\alpha, \overline{\mathrm{M}}_\beta)$ (see, for example, [6]).

Statistical characteristics of the second type define the microstructure of copolymer chains. The best known characteristics in this category are the fractions $\mathrm{P}\{U_k\}$ (probabilities) of sequences $\{U_k\}$ involving k monomeric units. The simplest among them are the dyads $\{U_2\}$, the complete set of which, for example, for a binary copolymer is composed of four pairs of monomeric units $\overline{\mathrm{M}}_1\overline{\mathrm{M}}_1$, $\overline{\mathrm{M}}_1\overline{\mathrm{M}}_2$, $\overline{\mathrm{M}}_2\overline{\mathrm{M}}_1$, $\overline{\mathrm{M}}_2\overline{\mathrm{M}}_2$. The number of the types of k-ad in chains of m-component copolymers grows exponentially as m^k so that with practical purposes in mind it is generally enough to restrict the consideration to sequences $\{U_k\}$ with moderate values of k. Their calculation turns out to be rather useful when carrying out mathematical modeling of the copolymerization processes for two reasons.

Firstly, some of the important characteristics of the performance properties of copolymers can be expressed through the fractions of these sequences. So, for example, dealing with copolymers they normally make extensive use of semiempirical relationships relating the glass transition temperature, T_g, with the fractions of dyads $P\{\overline{M}_\alpha \overline{M}_\beta\}$ in macromolecules [7]. The simplest of such relationships reads:

$$T_g = \sum_{\alpha\beta}^{m} T_{\alpha\beta}\, P\left\{\overline{M}_\alpha \overline{M}_\beta\right\} \tag{5}$$

where $T_{\alpha\alpha}$ and $T_{\alpha\beta}$ ($\alpha \neq \beta$) are, respectively, the glass transition temperature of a homopolymer with units $\overline{M}_\alpha$ and a regular alternating binary copolymer containing units $\overline{M}_\alpha$ and $\overline{M}_\beta$. Secondly, the probabilities $P\{U_k\}$ can be determined for k=2–4 with high accuracy by means of spectroscopic methods [8, 9]. A comparison of these experimental data with the results of mathematical modeling performed in the framework of the chosen kinetic model allows conclusions to be drawn about its adequacy for the description of the process of interest.

As well as the isomerism of copolymers that is due to differences in fashions of alternation of monomeric units along their chains, other types of isomerism exist [10]. These latter can show up even in homopolymer molecules provided several ways are conceivable for a monomer to enter into a polymer chain in the course of its propagation. So, asymmetric monomeric units can be arranged in macromolecules following "head-to-tail" or "head-to-head" - "tail-to-tail" patterns. As well as such a structural isomerism, in certain polymers there is also a possibility of stereoisomerism. Some isomers are known to vary substantially in their service properties. This circumstance should always be allowed for when selecting an appropriate kinetic model of polymerization. Ways to do this are analogous to that considered above with the only distinction being a more general specification of the set of transient states of the stochastic process of conventional movement along a polymer chain.

Frequently, the microstructure of polymer chains can be conveniently characterized by parameters representing some combinations of probabilities $P\{U_k\}$ rather than by the probabilities themselves. For a binary copolymer it is suggested [11] that as such a parameter the microheterogeneity coefficient $K_M = P\{\overline{M}_1\overline{M}_2\}/P\{\overline{M}_1\}P\{\overline{M}_2\}$ is employed. In the case of copolymers comprising long blocks of both monomeric units the value of K_M is close to zero, whereas for a regularly alternating copolymer K_M=2. Between these two extreme cases, corresponding to absolutely ordered arrangement of monomeric units in copolymer macromolecules, there lay all possible values of the microheterogeneity coefficient. To the most disordered arrangement of units, characterized by the Bernoullian random process, the value K_M=1 corresponds. Consequently, proceeding from the absolute value of the deviation of K_M from unity it is possible to make a conclusion about the degree of ordering of sequence distribution in chains of binary copolymers. On the other hand the direction of such a deviation testifies to the tendency of monomeric units either to regular alternation (if $K_M>1$) or to the formation of long blocks of both types of units (if $K_M<1$).

The microheterogeneity coefficient was introduced only for the description of the microstructure of binary copolymers with symmetric units. At increased number of unit types and/or when account is taken of structural isomerism, the role of K_M will be performed by other parameters analogous to it. A general strategy for the choice of these latter has been elaborated in detail [12], while their values have been measured via NMR spectroscopic techniques for a variety of polycondensation polymers [13].

It should be emphasized that for Markovian copolymers a knowledge of the values of structural parameters of such a kind will suffice to find the probability of any sequence $\{U_k\}$, i.e. for an exhaustive description of the microstructure of the chains of these copolymers with a given average composition. As for the composition distribution of Markovian copolymers, this obeys for any fraction of l-mers the Gaussian formula whose covariance matrix elements are $D_{\alpha\beta}/l$ where $D_{\alpha\beta}$ depend solely on the values of structural parameters [2]. The calculation of their dependence on time, and the stoichiometric and kinetic parameters of the reaction system permits a complete statistical description of the chemical structure of Markovian copolymers to be accomplished. The above reasoning reveals to which extent the mathematical modeling of the processes of the copolymer synthesis is easier to perform provided the alternation of units in macromolecules is known to obey Markovian statistics.

One more quantitative way to characterize the chemical structure of copolymers is based on the consideration of chemical correlation functions (correlators) [2]. The simplest of these, $Y_{\alpha\beta}(k)$, describes the joint probability of finding two randomly chosen monomeric units divided along the macromolecule by an arbitrary sequence U_k:

$$Y_{\alpha\beta}(k)=\sum_{U_k} P\left\{\overline{\mathrm{M}}_\alpha U_k \overline{\mathrm{M}}_\beta\right\} \tag{6}$$

The correlator (6) is of the utmost importance because its generating function enters into an expression which describes the angular dependence of intensity of scattering of light or neutrons [3]. It is natural to extend expression (6) for the two-point chemical correlation function by introducing the n-point correlator $Y_{\alpha_1 \ldots \alpha_n}(k_1 \ldots, k_{n-1})$ which equals the joint probability of finding in a macromolecule n monomeric units $\overline{M}_{\alpha_1}, \ldots, \overline{M}_{\alpha_n}$ divided by $(n-1)$ arbitrary sequences $U_{k_1}, \ldots, U_{k_{n-1}}$ comprising $k_1, \ldots, k_{n-1}$ units, respectively [14]. Generating functions of such correlators play a key role in the theory of microphase separation and in the statistical thermodynamics of block copolymers [3].

2
Peculiarities of the Description of Branched Polymers

Up to this point only polymers of linear structure have been addressed. However, a great many polymers exist whose molecules exhibit other topological configurations. The best known of these are comb-like, star-like, tree-like and network polymers. An inherent feature of all of them is the presence of branching units and polymer chains adjacent to them. Any molecule of such polymers may be depicted as a molecular graph with vertices and edges corresponding, respectively, to branching units and polymer chains. If these latter contain monomeric units of identical type they can be marked by assigning to every edge of a molecular graph the number equal to the length of the corresponding chain. To each macromolecule of a branched or crosslinked homopolymer there is a corresponding marked molecular graph, whereas a macroscopic polymer specimen is represented by an ensemble of such graphs. An exhaustive statistical description of this specimen can be realized by specifying the probability measure on the set of marked molecular graphs. For star-like and comb-like polymers finding the above probability distribution is not a matter of serious concern because the topology of their molecules is fixed. A completely different situation occurs for statistically branched tree-like and crosslinked polymers where the topology of molecules is stochastic. When calculating average geometrical sizes of macromolecules of such polymers, their optical properties, viscosity and other characteristics, it is necessary to perform an averaging over the topological configurations of the macromolecules.

An attempt to construct the probability measure on a set of such configurations has been successful for some classes of statistically branched polymers whose molecules do not comprise cycles. For each of them there is a corresponding molecular graph, referred to as a tree, while for a polymer specimen as a whole there is a set of such trees, which has been designated as a molecular forest. The latter may be transformed into a forest of rooted trees which are obtained from molecular unrooted trees by means of the successive choice of every vertex as a root. Such a transformation retains the probability measure, so that the only characteristic to determine is the probability distribution of the rooted trees. Each of them may, in turn, be envisaged as a genealogical tree, describing the history of a family or some realization of a stochastic branching process of birth and death of particles. The most simple of these is the Galton-Watson process, where the probability distribution for a particle to give birth to a certain number of descendants is the same for every generation and does not depend on other particles. Gordon was the first [15] to discover that for some polycondensation branched polymers (which will be referred to here as Gordonian polymers) the probability distribution of rooted trees is characterized by the probability measure on a set of genealogical trees which are realizations of the Galton-Watson branching process. The theory of these stochastic processes has been elaborated in minute detail [16]. This enables any statistical characteristics of a Gordonian polymer to be expressed in a relatively easy way through the probability parameters of the corresponding branching processes. The only thing that remains to be established is the dependency of these parameters on time, on the constants of the elementary reactions and on the composition of the initial monomer mixture. Equations which permit the determination of these dependencies have been derived for a variety of Gordonian polymers [2]. The niche these latter occupy among statistically branched polymers obviously resembles that of Markovian copolymers among linear copolymers.

A simple algorithm [17] makes it possible to find the probability of any fragment of macromolecules of Gordonian polymers. Comparison of these probabilities with the data obtained by NMR spectroscopy provides the possibility to evaluate the adequacy of a chosen kinetic model of a synthesis process of a particular polymer specimen. The above-mentioned probabilities are also involved in the expressions for the glass transition temperature and some structure-additive properties of branched polymers [18, 19].

One special feature of the process of branched polymer formation is the possibility of the appearance in the reaction system of a macromolecule whose size is comparable with that of the reactor. The specificity of such a phenomenon, termed gelation, consists in the fact that the conversion range in which a drastic rise in viscosity takes place is very narrow. This gives grounds to deem that gelation occurs at a certain instant, called the gel-point. A later gradual increase in the weight fraction of gel ω_g takes place in the course of the process accompanied by a simultaneous decrease in the weight fraction of sol molecules ω_s. Tasks of crucial importance for the mathematical modeling of branched polymerization and polycondensation, as well as that of the processes of crosslinking of linear macromolecules, are the calculation of the gel-point and the determination of the dependence of $\omega_g=1-\omega_s$ on time. If monomeric units of several types are involved in the process it is necessary to determine analogous dependencies for composition of sol and gel. The latter represents a polymer network showing elastic properties. In line with current theoretical concepts [20], the elasticity modulus of this network is described by the cyclic rank of its molecular graph. It is equal, by definition, to the minimum number of edges that need to be removed to transform this cyclic graph into a tree-like one. Obviously, the determination of cyclic rank, pertaining to the global characteristics of the topological structure of the polymer network, constitutes one of the major challenges of mathematical modeling of network formation processes. This problem, like all the above-mentioned ones which are connected with the calculation of statistical characteristics of Gordonian polymers, can be solved using the mathematical apparatus of the theory of branching processes [2, 17, 21].

3 Kinetic Models of Macromolecular Reactions

When deriving material balance equations the rate of each component transformation in the reactor obeys the law of mass action. However, as distinct from the reactions with participation of exclusively low molecular weight substances, the number of such components in a polymer system and, consequently, the number of equations for them is practically infinite. The same is true for the constants of the rate of the reactions between individual components. The calculation of such a system becomes possible due to the general principle which underlies the description of the kinetics of many macromolecular reactions. It will be discussed in more detail below.

The processes of formation and chemical modification of polymer molecules are known to proceed as a result of chemical reactions of their reactive centers. The role of these latter can be performed by functional groups, free valence in radicals, double bonds and so on. Often the reactivity of the reactive center in a polymer molecule can be believed to be independent of its configuration (i.e., molecular weight, composition, chemical structure) as well as of the location of this center inside a macromolecule. This fundamental principle, put forward by Flory [22] and bearing his name, permits the use of the single rate constant of the elementary reaction of any pair of reactive centers of a particular type. The number of such elementary reactions in a polymer synthesis process is normally small enough to make it possible to characterize it using only a few rate constants. Thus, for instance, free-radical polymerization may be characterized by the rate constants of the elementary reactions of the initiation, propagation, termination and chain transfer.

The Flory principle allows a simple relationship between the rate constants of macromolecular reactions (whose number is infinite) with the corresponding rate constants of elementary reactions. According to this principle all chemically identical reactive centers are kinetically indistinguishable, so that the rate constant of the reaction between any two molecules is proportional to that of the elementary reaction between their reactive centers and to the numbers of these centers in reacting molecules. Therefore, the material balance equations will comprise as kinetic parameters the rate constants of only elementary reactions whose number is normally rather small.

The Flory principle is one of two assumptions underlying an *ideal* kinetic model of any process of the synthesis or chemical modification of polymers. The second assumption is associated with ignoring any reactions between reactive centers belonging to one and the same molecule. Clearly, in the absence of such intramolecular reactions, molecular graphs of all the components of a reaction system will contain no cycles. The last affirmation concerns sol molecules only. As for the gel the cyclization reaction between reactive centers of a polymer network is quite admissible in the framework of an ideal model.

The Flory principle, whose validity has been verified for a wide range of macromolecular reactions, is a good approximation for the description of the kinetics of many processes of polymer preparation. However, strong experimental evidence for the violation of this principle is currently available from the study of a number of macromolecular reactions. Possible reasons for such a violation may be connected with either *short-range* or *long-range* effects.

To the first of these pertain steric, inductive, catalytic or other effects responsible for the influence of reacted centers on the reactivity of neighboring centers of the same molecule. To take such substitution effects into account it has been suggested [2] to resort to the extended Flory principle. In accordance with the latter the reactivity of any reactive center depends exclusively on the local chemical structure of the fragment of the molecule containing this center, and it undergoes changes in the course of the process due to the evolution of this local structure induced by entering into the reaction of neighboring reactive centers. The influence of other factors on the reactivity is supposed here to be negligible. Within the framework of models of such a kind it has been suggested that small fragments of a molecule comprising the reactive centres should be chosen as kinetically independent elements, rather than these centres themselves (as is customary for the case of the ideal model). Using such an approach, generally applied for mathematical modeling of the processes of copolymerization, copolycondensation and polymer-analogous transformation, the number of parameters of a kinetic model still remains reasonable.

The specificity of the polymer nature of the reagents is the most strongly pronounced in the long-range effects where the reactivity of a reactive center is affected by the fragments positioned all over the molecule. Here mention should be made of intramolecular catalysis by functional groups spaced apart from the reactive center along the macromolecule but falling into its vicinity by virtue of the spatial conformational rearrangements of the polymer chain. The reactivity of its terminal reactive center under such effects may be controlled by the degree of polymerization of this macromolecule and the distribution along it of units containing catalytic groups. In case of fast reactions, like the recombination of macroradicals, the dependence of the rate constant on their lengths is governed by diffusion factors. Thermodynamic factors may also be responsible for the dependence of the rate of chemical reactions of macroradicals or macromolecules with low molecular weight compounds on their length and composition. The reason for such an effect has to do with the distinction of concentrations of low molecular weight reagents inside and outside the polymer chain due to preferential sorption of these reagents inside the polymer coil or globula. Although the allowance for long-range effects in the kinetic model makes the calculations far more complicated, this is sometimes quite indispensable in order to interpret some nontrivial effects revealed experimentally.

4 Methods of Calculation

When calculating the statistical characteristics of the primary structure of macromolecules, there is a need to tackle the problems of finding their average molecular weight, composition, molecular-weight and size-composition distribution as well as microstructural characteristics. To cope with this task two different approaches are applied: kinetic and statistical. The first approach implies setting up and solving the material balance equations for the concentrations of molecules of all types involved in the process of interest. Under the second approach every macromolecule is thought of as an individual realization of a particular stochastic process of conventional movement along the polymer chain. The probability of this realization equals the fraction of corresponding molecules among all those present in the reactor.

The statistical as well as the kinetic method of calculation of the parameters of the primary structure of polymers has advantages and disadvantages. The statistical method for the quantitative description of the products of macromolecular reactions was introduced more than fifty years ago by Flory [22] and since then it is widely recognized in the mathematical modeling of a variety of processes of synthesis and chemical modifications of polymers. The indisputable advantage of a statistical method consists of its ability to use a simple mathematical formalism to describe exhaustively the chemical structure of macromolecules in terms of just a few probability parameters. However the choice of an appropriate stochastic process for the probabilistic description of the specimen obtained under particular synthesis conditions cannot be made in principle within the framework of a very statistical method. Such a problem can be settled only by means of a kinetic approach. Moreover, recourse to the latter makes it possible to express the probability parameters of this stochastic process through the rate constants, concentrations of the reagents and other quantities characterizing the reaction system.

The kinetic method was found to be especially advantageous for mathematical modeling of the polymerization or polycondensation processes, described by the ideal kinetic model. In this case the material balance equations for the concentrations of the polymer molecules with a specified numbers of monomeric units and reactive centers may often be integrated analytically via the method of generating functions [2]. Such a function is entirely equivalent to the distribution of concentrations of molecules for size, composition and functionality. These concentrations are obtainable as coefficients of the expansion of generating function into power series. Recourse to this function is especially appealing if it is necessary to calculate statistical moments of the above distribution, which can be expressed through the derivatives of generating function at one point, namely, when the value of all its arguments is unity. An infinite set of ordinary differential equations for the distribution of concentrations of polymer molecules in the case of an ideal kinetic model is reducible to a single equation for generating function. This will be an ordinary differential equation or a first-order partial differential equation depending on whether polymers formed in the course of the synthesis process are linear or branched, respectively. The coefficients of this equation are controlled along with the rate constants of elementary reactions also by average concentrations of reactive centers and low molecular weight reagents. These concentrations can be found from the solution of a set of ordinary differential equations which is always closed provided the ideal kinetic model is applicable. In the framework of this model the elementary reactions can be considered separately from the reactions with participation of macromolecules. In many cases of practical interest partial differential equation for generating function can be integrated analytically using the method of characteristics.

When there is a need to calculate only the statistical moments of the distribution of molecules for size and composition, rather than to find the very distribution, the task becomes essentially easier. The fact is that for the processes of polymer synthesis which may be described by the ideal kinetic model the set of equations for the statistical moments is always closed.

This closure property is also inherent to a set of differential equations for arbitrary sequences $\{U_k\}$ in macromolecules of linear copolymers as well as for analogous fragments in branched polymers. Hence, in principle, the kinetic method enables the determination of statistical characteristics of the chemical structure of noncyclic polymers, provided the Flory principle holds for all the chemical reactions involved in their synthesis. It is essential here that the Flory principle is meant not in its original version but in the extended one [2]. Hence under mathematical modeling the employment of the kinetic models of macromolecular reactions where the violation of ideality is connected only with the short-range effects will not create new fundamental problems as compared with ideal models.

Another kind of situation arises when it is necessary to take into account the long-range effects. Here, as a rule, attempts to obtain analytical results have not met with success. Unlike the case of the ideal model the equations for statistical moments of distribution of polymers for size and composition as well as for the fractions of the fragments of macromolecules turn out normally to be unclosed. Consequently, to determine the above statistical characteristics, the necessity arises for a numerical solution to the material balance equations for the concentration of molecules with a fixed number of monomeric units and reactive centers. The difficulties in solving the infinite set of ordinary differential equations emerging here can be obviated by switching from discrete variables, characterizing macromolecule size and composition, to continuous ones. In this case the mathematical problem may be reduced to the solution of one or several partial differential equations.

Traditionally, to calculate the different characteristics of the molecular structure of polymers in macromolecular chemistry, a statistical approach has been used. In line with the most general formulation of this approach to a set of polymer molecules in a particular ensemble it has been suggested that a statistically equivalent set of realizations of a certain stochastic process be placed in correspondence. Instead of averaging over the ensemble of molecules for the calculation of statistical characteristics of a polymer an analogous averaging was recommended to perform over realizations of the corresponding stochastic process. Methods of such averaging have been thoroughly elaborated for some stochastic processes. That is why the procedure of finding the statistical characteristics of the chemical structure of Markovian and Gordonian polymers is a relatively easy task.

For a number of copolymers, whose kinetics of formation is described by nonideal models, the statistics of alternation of monomeric units in macromolecules cannot be characterized by a Markov chain; however, it may be reduced to the extended Markov chain provided that units apart from their chemical nature are differentiated by some additional attributes. These latter can vary for different kinetic models. Thus, for instance, in the penultimate model of free-radical copolymerization [23] such an attribute is the type of preceding unit. One more type of extended Markov chain characterizes the sequence distribution in products of complex-radical copolymerization [24].

To calculate the characteristics of a copolymer chemical structure in the cases discussed above using a statistical method, the following strategy can be adopted. Firstly, it is necessary to change from an ordinary (non-Markovian!) chain to the extended one, marking monomeric units by labels in accordance with a certain attribute. Thereafter, because the extended chain is Markovian, it is possible to derive the expressions for the statistical characteristics of the ensemble of macromolecules with labeled units. To eliminate superfluous information contained in these expressions, one should "erase" the labels which the monomeric units were marked with. The erasing procedure implies here the summing of the derived expressions over corresponding indices, characterizing the state of a monomeric unit of fixed chemical structure. This procedure is conditioned by the fact that every state of an ordinary chain is actually the sum of several states of the extended Markov chain over which the mentioned summing is carried out.

Under current treatment of statistical method a set of the states of the Markovian stochastic process describing the ensemble of macromolecules with labeled units can be not only discrete but also continuous. So, for instance, when the description of the products of "living" anionic copolymerization is performed within the framework of a terminal model the role of the label characterizing the state of a monomeric unit is played by the moment when this unit forms in the course of a macroradical growth [25].

The above reasoning concerning the statistical method may also be extended to statistically branched polymers. Very often, when their formation occurs under the condition of the applicability of the ideal kinetic model, it is possible to calculate the characteristics of the molecular structure of such Gordonian polymers proceeding from the equations of the Galton-Watson branching process. The role of reproducing particles for this stochastic process in the case of polycondensation polymers is performed by reacted functional groups of different types. Within the framework of kinetic models allowing for the "substitution effect" these groups should, generally speaking, be discriminated on the basis of both the instant of their entering into the reaction and the instantaneous configuration of a monomeric unit to which this group pertains [26]. The probability measure on the set of sol molecules which are formed during such a branched polycondensation of monomers with kinetically dependent functional groups, instead of the Galton-Watson process, will be specified by the *general* branching process. The latter is defined as a certain extended Markovian random process whose mathematical theory is now sufficiently developed [16].

It should be pointed out that the employment of the statistical method in all its versions is, in essence, of a purely formal character. The expressions for statistical characteristics of polymers are traditionally obtained via speculative probabilistic considerations whose correctness is to a great extent predetermined by the author's scientific intuition. Naturally, remaining within the framework of such a formal statistical approach, it is impossible in principle to establish a rigorous correspondence between a kinetic model of the reactions of polymer synthesis and the type of stochastic process which adequately describes the statistical characteristics of macromolecules formed. This is the reason why the question concerning the areas of applicability of a statistical approach often remains unanswered. With this in mind a general concept has been developed [2] consisting in a rigorous (in the framework of models generally accepted today in polymer chemistry) substantiation of the statistical method for a variety of classes of the processes of polymer synthesis on the basis of their kinetic consideration. The expressions for the distribution of macromolecules for size and composition arrived at by means of such a consideration are compared with analogous ones but derived proceeding from the theory of stochastic processes. The coincidence of the expressions for these distributions testifies to the correctness of the chosen version of the statistical approach. Under such a comparison the dependencies are simultaneously established of probability parameters of a stochastic process on stoichiometric and kinetic parameters of the reaction system. Then the statistical characteristics of interest, prescribed by the structure of polymer molecules, can be calculated by the statistical method. This general concept has been adopted for mathematical modeling of different classes of polymers [2, 27, 28].

In subsequent sections some results will be reported relevant to the theoretical consideration of several principle processes of the synthesis of polymers described by various kinetic models. This information may be useful in gaining a better understanding of the potentialities of the statistical chemistry of polymers.

5
Chain Polymerization

The role of reactive centers is performed here by free radicals or ions whose reaction with double bonds in monomer molecules leads to the growth of a polymer chain. The time of its formation may be either essentially less than that of monomer consumption or comparable with it. The first case takes place in the processes of free-radical polymerization whereas the second one is peculiar to the processes of "living" anionic polymerization. The distinction between these two cases is the most greatly pronounced under copolymerization of two and more monomers when the change in their concentrations over the course of the synthesis induces chemical inhomogeneity of the products formed not only for size but for composition as well.

An individual approach has been developed which provides a quantitative description of living free-radical copolymerizations which are presently of utmost academic and industrial interest (see, for example, [29–31] and references cited therein). A feature peculiar to these processes is the stepwise growth of a macromolecule. Every step represents here an elementary chain formed at some invariable monomer mixture composition. As for the polymer chain this is composed of a random number of elementary chains formed at different times. Evidently, the statistical analysis of the ensemble of such polymer chains [32] presents a far more complex theoretical problem when compared with both traditional free-radical copolymerization and anionic copolymerization. The consideration of living free-radical copolymerization is beyond the scope of the present review.

5.1 Free-Radical Copolymerization

Monomer concentrations $M_\alpha(\alpha=1,\ldots,m)$ in a reaction system have no time to alter during the period of formation of every macromolecule so that the propagation of any copolymer chain occurs under fixed external conditions. This permits one to calculate the statistical characteristics of the products of copolymerization under specified values $\{M_\alpha\}$ and then to average all these "instantaneous" characteristics with allowance for the drift of monomer concentrations during the synthesis. Such a two-stage procedure of calculation, where first statistical problems are solved before dealing with dynamic ones, is exclusively predetermined by the very specificity of free-radical copolymerization and does not depend on the kinetic model chosen. The latter gives the explicit dependencies of the "instantaneous" statistical characteristics on monomers' concentrations and the rate constants of the elementary reactions.

5.1.1 *Ideal (Ultimate) Model*

In the framework of this ultimate model [33] there are m^2 constants of the rate of the chain propagation $k_{\alpha\beta}$ describing the addition of monomer M_β to the radical R_α whose reactivity is controlled solely by the type α of its terminal unit. Elementary reactions of chain termination due to chemical interaction of radicals R_α and R_β is characterized by m^2 kinetic parameters $k^t_{\alpha\beta}$. The stochastic process describing macromolecules, formed at any moment in time t, is a Markov chain with transition matrix whose elements are expressed through the concentrations R_α and M_α of radicals and monomers at this particular moment in the following way [1, 34]:

$$\nu_{\alpha\beta}=k_{\alpha\beta}M_\beta\left(\sum_{\gamma=1}^{m}k_{\alpha\gamma}M_\gamma+T_\alpha\right)^{-1}\quad(1\le\alpha,\beta\le m),\quad T_\alpha\equiv\sum_{\gamma=1}^{m}k^t_{\alpha\gamma}R_\gamma$$
$$\nu_{00}=1,\quad \nu_{0\beta}=0,\quad \nu_{\alpha 0}=T_\alpha\left(\sum_{\gamma=1}^{m}k_{\alpha\gamma}M_\gamma+T_\alpha\right)^{-1}\qquad(7)$$

When calculating the average copolymer composition and the probabilities $P\{U_k\}$ of the sequences of monomeric units it is possible to set $T_\alpha=0$ in the expressions in (7), that is to neglect the finiteness of the size of the macromolecules. In this case the absorbing Markov chain (7) is replaced by the ergodic Markov chain with transition matrix **Q** whose elements:

$$v_{\alpha\beta} = a_{\alpha\beta} x_\beta \left(\sum_{\gamma=1}^{m} a_{\alpha\gamma} x_\gamma \right)^{-1}, \text{where } a_{\alpha\beta} \equiv \frac{k_{\alpha\beta}}{k_{\alpha\alpha}} \equiv \frac{1}{r_{\alpha\beta}} \quad (1 \le \alpha, \beta \le m) \tag{8}$$

are governed along with molar fractions x_α of monomers in the reaction system only by the values of the the reactivity ratios $r_{\alpha\beta}$.

These latter have been tabulated for a great many pairs (M_α,M_β) of particular monomers [35]. This circumstance offers in many cases an opportunity to carry out the calculation of statistical characteristics of the products of multicomponent copolymerization without the need for additional experiments to determine the kinetic parameters of the model.

The instantaneous composition of a copolymer **X** formed at a monomer mixture composition **x** coincides, provided the ideal model is applicable, with stationary vector π of matrix **Q** with the elements (8). The mathematical apparatus of the theory of Markov chains permits immediately one to wright out of the expression for the probability of any sequence $P\{U_k\}$ in macromolecules formed at given **x**. This provides an exhaustive solution to the problem of sequence distribution for copolymers synthesized at initial conversions $p<<1$ when the monomer mixture composition **x** has had no time to deviate noticeably from its initial value $\mathbf{x}^0$. As for the high-conversion copolymerization products they evidently represent a mixture of Markovian copolymers prepared at different times, i.e. under different concentrations of monomers in the reaction system. Consequently, in order to calculate the probability of a certain sequence U_k, it is necessary to average its instantaneous value $P\{U_k\}$ over all conversions p' preceding the conversion p up to which the synthesis was conducted.

$$\langle P\{U_k\} \rangle \equiv \frac{1}{p} \int_0^p P\{U_k\} dp' \tag{9}$$

Because the dependence of probability $P\{U_k\}$ on **x** should be established by means of the theory of Markov chains, in order to make such an averaging it is necessary to know how the monomer mixture composition drifts with conversion. This kind of information is available [2, 27] from the solution of the following set of differential equations:

$$(1-p)\frac{dx_\alpha}{dp} = x_\alpha - X_\alpha(\mathbf{x}) \quad (\alpha = 1,\dots,m) \tag{10}$$

which should be found under the initial condition $\mathbf{x}=\mathbf{x}^0$ at $p=0$.

Noteworthy, when calculating composition and sequence distribution in copolymers, only the knowledge of the reactivity ratio $r_{\alpha\beta}$ which characterizes chain propagation reactions is imperative. However these kinetic parameters turn out to be insufficient to determine the SCD which, according to expression (1), represents the product of two distributions. The first of them, SD, is controlled, along with the initial concentrations of the monomers and the initiator, also by the rate constants of all the elementary reactions of initiation, propagation, termination and chain transfer. The second distribution, CD, at any fixed composition of monomer mixture **x** is defined by the Gauss formula with the center **X** and covariance matrix having elements $D_{\alpha\beta}/l$. The quantities X_α and $D_{\alpha\beta}$ are governed exclusively by **x** and the reactivity ratios $\{r_{\alpha\beta}\}$. Dependence on these parameters has been established [2, 27] for the copolymerization of an arbitrary number of monomer types m proceeding from the theory of the ergodic Markov chains with transition matrix (8). To describe the phase behavior of melts of such ergodic copolymers, a phase diagram was constructed and its dependence on thermodynamic and structural parameters analyzed [4].

The composition inhomogeneity of ergodic copolymers, which are the products of low-conversion copolymerization, is predetermined only by the stochastic character of the addition of monomers to the growing macroradical, since the concentration of monomers during its lifetime remains practically constant. In parallel with this instantaneous component of the composition inhomogeneity there exists another induced by the drift of composition of monomers in the course of the synthesis. The second component of the CD, referred to as the conversional component, is known to make a far more substantial contribution in comparison with the first component to composition inhomogeneity of the products of high-conversion copolymerization. This is due to the fact that the elements of composition covariance matrix characterizing the width of instantaneous Gaussian distribution are reciprocal to size l of macromolecules. Within the limit $l \to \infty$ the instantaneous component of composition distribution can be completely neglected, unlike its conversional component. Mathematically this means that the distribution $W(l/\zeta)$ (1) does not depend on l and is equal to the Dirak delta-function $\delta(\zeta - \mathbf{X})$ which is distinct from zero only at $\zeta = \mathbf{X}$. In other words all macromolecules formed under fixed monomer mixture composition **x** should have identical composition **X**. Within the framework of this approximation the CD of the copolymerization products, present in the reaction system under conversion p, is not controlled by l and is prescribed apart from the stoichiometry of the initial monomer mixture $\mathbf{x}^0$ only by the value of the reactivity ratios $\{\mathbf{r}_{\alpha\beta}\}$. Since these products represent a mixture of Markovian copolymers, their CD $\langle W(l/\zeta)\rangle$ can be obtained, in line with formula (9), by averaging the instantaneous CD over conversions $p'<p$. In practice it is quite sufficient to use the set of one-dimensional partial distributions of each α-th component

$$\left\langle \delta\left(\zeta_\alpha - X_\alpha\right)\right\rangle \equiv \frac{1}{p}\int_0^p \delta\left(\zeta_\alpha - X_\alpha\right)dp' = \frac{1}{p}\sum_{p'}\left|\frac{dX_\alpha}{dp}\right|^{-1} \tag{11}$$

where the sum is over those values of conversion $p=p'$ at which the component X_α of instantaneous copolymer composition vector **X** is equal to ζ_α. Knowing the solution **x**(p) of Eq. (10) it becomes possible to find, proceeding from the known dependence of **X** on **x**, distribution (11), as well as the elements of covariance matrix of the m-component copolymer

$$\lambda_{\alpha\beta} = \langle X_\alpha X_\beta \rangle - \langle X_\alpha \rangle \langle X_\beta \rangle \tag{12}$$

obtained at arbitrary values of conversion p. The only kinetic parameters which must be known for the determination of the above-mentioned statistical characteristics of a copolymer are the values of the reactivity ratios $\{r_{\alpha\beta}\}$.

Of major interest in practice are copolymers prepared under high conversions. They represent a mixture of Markovian copolymers formed under different conditions. To calculate the statistical characteristics of these mixtures the averaging (9, 11) should be carried out of their instantaneous values over conversions. The results of such an averaging strongly depend on the character of the evolution of monomer mixture composition during the synthesis. This evolution is described by the solutions of the set of Eqs. (10). To every such a solution under m-component copolymerization a trajectory **x**(p) corresponds inside the m-simplex $x_1+\ldots+x_m=1$ which is an interval of unit length (m=2), equilateral triangle (m=3), regular tetrahedron (m=4) and so on. The trajectory x(p) begins at the point $\mathbf{x}(0)=\mathbf{x}^0$ and terminates at the point $\mathbf{x}(1)=\mathbf{x}^*$, termed attractor. Vector $\mathbf{x}^*$ may either have all components but one equal to zero, i.e. correspond to a homopolymer, or have more than one non-zero component corresponding to the azeotropic copolymer. Its instantaneous composition **X** does not alter with conversion and coincides with the monomer mixture composition **x**. At certain sets of kinetic parameters $\{r_{\alpha\beta}\}$ a system may have several attractors. Each of them possesses its own basin of attraction; if the initial point of the trajectory falls within this basin, this ensures the termination of the latter just in this very attractor. If two initial compositions of monomer mixture positioned close to one another within the m-simplex are selected, but which fall into the basins of attraction of different attractors, then the copolymer specimens obtained under complete conversion will have practically identical average compositions but will differ markedly in their composition distribution. The reason for such a discrepancy should be attributed to the qualitatively distinct character of the evolution with conversion of instantaneous copolymer composition **X** which is responsible for the shape of this distribution (11). The latter may be narrow in one of the specimens and wide in the other.

Such a distinction in composition inhomogeneity of the copolymers may have caused the variation in transparency which was observed experimentally by Slocombe [36]. He examined forty-four three-component systems and established several empirical rules enabling the interpretation of experimental data on the transparency of high-conversion terpolymers. These empirical rules were explained later [37] in terms of the theory of dynamic systems whose methods have been successfully employed for qualitative analysis of the solutions of the set of dynamic equations of multicomponent copolymerization (10). Using these methods it has been possible not only to substantiate the empirical rules put forward by Slocombe but also to predict the transparency regions of the products of terpolymerization of styrene with some acrylates and methyl methacrylate [37, 38].

From the above reasoning it may be concluded that the quantitative theory as it stands today gives the opportunity to provide an exhaustive description of the chemical structure of the products of free-radical copolymerization of any number of monomers m.

A user-friendly computer program has been developed (A.S.Yakovlev, S.I.Kuchanov "Copolymerization for Windows") which makes it possible at any values of conversion to calculate for m=2–6 along with the composition of monomer mixture $\mathbf{x}$, such statistical characteristics as instantaneous $\mathbf{X}$ and average $\langle \mathbf{X} \rangle$ copolymer composition as well as the fractions $\langle P\{U_k\} \rangle$ of sequences U_k with k=2–4 and composition distribution. To calculate all these characteristics at any composition of initial monomer mixture $\mathbf{x}^0$ it is only necessary to have as input parameters the values of the reactivity ratios $\{\mathbf{r}_{\alpha\beta}\}$. In addition, for the purpose of the prediction of the transparency and heat resistance of the products of copolymerization, provision is made in the program for the calculation of their enthalpy of mixing (4) and glass transition temperature (5). Having completed the discussion of the ideal model we will now switch to more complicated kinetic models.

5.1.2
Penultimate Model

This is the simplest of the models where violation of the Flory principle is permitted. The assumption behind this model stipulates that the reactivity of a polymer radical is predetermined by the type of both its ultimate and penultimate units [23]. Here, the pairs of terminal units $\sim\overline{\mathrm{M}}_\alpha\overline{\mathrm{M}}_\beta$ act, along with monomers M_γ, as kinetically independent elements, so that there are m^3 constants of the rate of elementary reactions of chain propagation $k_{\alpha\beta\gamma}$. The stochastic process of conventional movement along macromolecules formed at fixed $\mathbf{x}$ will be Markovian, provided that monomeric units are differentiated by the type of preceding unit. In this case the number of transient states S_α of the extended Markov chain is m^2 in accordance with the number of pairs of monomeric units. No special problems presents writing down the elements of the matrix of the transitions Q of such a chain [1, 10, 34, 39] and deriving by means of the mathematical apparatus of the Markov chains the expressions for the instantaneous statistical characteristics of copolymers. By way of illustration this matrix will be presented for the case of binary copolymerization:

$$\mathbf{Q}=\begin{pmatrix} \nu_{11} & \nu_{12} & 0 & 0 \\ 0 & 0 & \nu_{23} & \nu_{24} \\ \nu_{31} & \nu_{32} & 0 & 0 \\ 0 & 0 & \nu_{43} & \nu_{44} \end{pmatrix} \tag{13}$$

where probabilities of the transitions $\nu_{\alpha\beta}\neq 0$ between irreversible states S_α:

$$S_1{\sim}\overline{M}_1\overline{M}_1,\ S_2{\sim}\overline{M}_1\overline{M}_2,\ S_3{\sim}\overline{M}_2\overline{M}_1,\ S_4{\sim}\overline{M}_2\overline{M}_2 \tag{14}$$

may be presented as follows:

$$\nu_{\alpha\beta}=\frac{r^{(\alpha)}x_1}{x_2+r^{(\alpha)}x_1}\ (\beta=1,3),\qquad \nu_{\alpha\beta}=\frac{x_2}{x_2+r^{(\alpha)}x_1}\ (\beta=2,4) \tag{15}$$

Four kinetic parameters $r^{(\alpha)}$ are related to the reactivity ratios:

$$r_1=\frac{k_{111}}{k_{112}},\quad r_1'=\frac{k_{211}}{k_{212}},\quad r_2=\frac{k_{222}}{k_{221}},\quad r_2'=\frac{k_{122}}{k_{121}} \tag{16}$$

by the simple relationships:

$$r^{(1)}=r_1,\quad r^{(2)}=1/r_2',\quad r^{(3)}=r_1',\quad r^{(4)}=1/r_2 \tag{17}$$

In order to obtain the expression for the components of the vector of instantaneous copolymer composition it is necessary, according to general algorithm, to firstly determine the stationary vector π of the extended Markov chain with the matrix of transitions (13) which describes the stochastic process of conventional movement along macromolecules with labeled units and then to erase the labels. In this particular case such a procedure reduces to the summation:

$$X_1=\pi_1+\pi_3 \qquad X_2=\pi_2+\pi_4 \tag{18}$$

corresponding to the enhancement of the states (14) via their coupling. Analogous erasing of the labels may be accomplished when calculating other statistical characteristics of copolymers. After determining their instantaneous values in this way, one should proceed further to conversional averaging identical to that performed under the description of copolymerization within the framework of the ideal model. This means that formulas (9)–(12) remain valid.

5.1.3
Complex Participation Model

Currently this model is one of the most commonly used in the theory of free-radical copolymerization. The formation of a donor–acceptor complex $M_\alpha\ldots M_\beta$ between monomers M_α and M_β in some systems is responsible for a number of peculiarities absent in the case of the ideal model. Such peculiarities are due to the fact that besides the single monomer addition to a propagating radical, a possibility also exists of monomer addition in pairs as a complex. Here the role of kinetically independent elements is played by ultimate units $\sim\overline{M}_\alpha$ of growing chains as well as by free (M_β) and complex-bound (M^*_β) monomers, whose constants of the rate of addition to the macroradical with α-th ultimate unit will be

$k_{\alpha\beta}$ and $k^*_{\alpha\beta}$, respectively. This kinetic model has been put forward [40] for the description of binary copolymerization. Here, for simplicity, we restrict ourselves to consideration of this simplest case in order to illustrate the ideas of general approach [24] to the description of copolymerization of any number of monomers.

Rigorous kinetic analysis has shown [41] that the products of binary copolymerization, formed under the conditions of constant concentrations of monomers, may be described by the extended Markov chain with four states S_α, if to label monomeric units conventionally coloring them in red and black. Unit $\overline{M}_\alpha$ is presumed to be black when the corresponding monomer M_α adds to the radical as the first monomer of the complex. In other cases, when monomer M_α adds individually or as the second monomer of the complex, the unit $\overline{M}_\alpha$ is assumed to be red. As a result the state of a monomeric unit is characterized by two attributes, one of which is its type (α=1,2) while the second one is its color (r,b). For example, we shall speak about the unit being in the state S_1 provided it is of the first type and red-colored, i.e. $\overline{M}^r{}_1$. The other states S_α are determined in a similar manner:

$$S_1 \sim \overline{M}^r_1,\ S_2 \sim \overline{M}^b_1,\ S_3 \sim \overline{M}^r_2,\ S_4 \sim \overline{M}^b_2 \tag{19}$$

Essentially, in realistic polymer chains, a monomeric unit does not "remember" the way it appeared in the macroradical. All the experimental characteristics of a copolymer chemical structure are naturally described in terms of uncolored units. Consequently, having preliminarily calculated these characteristics in the ensemble of macromolecules with colored units, it is then necessary to "erase" colors bearing in mind that every state in a chain of uncolored units is an enhancement of a corresponding pair of states in a chain of colored units. The latter is the Markov chain with transient states (19), whose matrix of transitions looks as follows:

$$\mathbf{Q} = \begin{pmatrix} v_{11} & v_{12} & v_{13} & v_{14} \\ 0 & 0 & 1 & 0 \\ v_{31} & v_{32} & v_{33} & v_{34} \\ 1 & 0 & 0 & 0 \end{pmatrix} \tag{20}$$

The elements of this matrix which differ from zero and unity

$$\begin{aligned} & v_{1\beta} = d_{1\beta}/d_1 \qquad v_{3\beta} = d_{3\beta}/d_3 \qquad (\beta = 1,2,3,4) \\ & d_{11} = M_1 \quad d_{12} = a^*_{11} M_{12} \quad d_{13} = a_{12} M_2 \quad d_{14} = a^*_{12} M_{12} \\ & d_{31} = a_{21} M_1 \quad d_{32} = a^*_{21} M_{12} \quad d_{33} = M_2 \quad d_{34} = a^*_{22} M_{12} \\ & d_\alpha = d_{\alpha 1} + d_{\alpha 2} + d_{\alpha 3} + d_{\alpha 4}\ (\alpha = 1,3), \quad a_{\alpha\beta} = 1/r_{\alpha\beta},\ a^*_{\alpha\beta} = 1/r^*_{\alpha\beta} \end{aligned} \tag{21}$$

contain as kinetic parameters, apart from $\{r_{\alpha\beta}\}$ which occur in the ideal model, another four dimensionless parameters:

$$r_{11}^{*}=\frac{k_{11}}{k_{11}^{*}} \qquad r_{12}^{*}=\frac{k_{11}}{k_{12}^{*}} \qquad r_{21}^{*}=\frac{k_{22}}{k_{21}^{*}} \qquad r_{22}^{*}=\frac{k_{22}}{k_{22}^{*}} \tag{22}$$

Upon expressing from the equilibrium condition the complex concentration M_{12} through the concentrations of monomers, and substituting the expression found into relationship (21) we obtain, invoking the formalism of the Markov chains, final formulas enabling us to calculate instantaneous statistical characteristics of the ensemble of macromolecules with colored units. A subsequent color erasing procedure is carried out in the manner described above. For example, when calculating instantaneous copolymer composition, this procedure corresponds to the summation of the appropriate components of the stationary vector π of the extended Markov chain:

$$X_1=\pi_1+\pi_2 \qquad X_2=\pi_3+\pi_4 \tag{23}$$

5.1.4
Preferential Sorption Model

During experimental investigation of the products of low-conversion binary copolymerization of some monomers a number of peculiarities were revealed [42–47] which could not be interpreted in terms of traditional kinetic models allowing for only short-range effects. So, the uniform composition copolymer specimens of styrene with methacrylic acid synthesized in carbon tetrachloride and dioxane solution exhibited identical sequence distribution regardless of the solvent type [42]. These results led to the conclusion that the same mechanism and the same set of reactivity ratios were applicable for both systems despite the substantial divergence in the shape of the curve $\mathbf{X}(\mathbf{x})$. Later, Harwood [43] extended this finding to other solvents and monomers. It was suggested [42, 43] that the main reason responsible for this effect is associated with the peculiarities caused by the partitioning of monomer molecules between the bulk of the reaction mixture and the domain close to the reactive center of the growing polymer chain. The above phenomenon, originating from the preferential sorption of the monomers in such a domain, has been named by Harwood [43] the "bootstrap effect" because of the capability of the macroradical to control its environment.

Some other anomalies, falling well outside the boundaries of traditional theories, have been experimentally found by Semchikov, Smirnova et al. (see [44–47] and references cited therein) who examined bulk low-conversion copolymerization of about thirty pairs constituted of the commonly encountered vinyl monomers. These anomalies can be listed as follows:

1. The composition of a relevant number of copolymers is tangibly controlled by their molecular weight.

2. Copolymers synthesized under conversion of less than ten percent exhibit composition heterogeneity substantially exceeding that predicted by the traditional theories.
3. Each copolymer chain is chemically inhomogeneous along its length.

In order to explain the above-mentioned anomalies an original model of the *m*-component copolymerization has been put forward [48], taking into account chemical factors along with the also thermodynamic ones. It is pertinent to dwell briefly on the key points of this model.

At the initial stage of bulk copolymerization the reaction system represents the diluted solution of macromolecules in monomers. Every radical here is an individual microreactor with boundaries permeable to monomer molecules, whose concentrations in this microreactor are governed by the thermodynamic equilibrium whereas the polymer chain propagation is kinetically controlled. The evolution of the composition of a macroradical **X** under the increase of its length l is described by the set of equations:

$$\frac{dX_\alpha}{d\xi} = \pi_\alpha(\mathbf{x}) - X_\alpha, \text{ where } \xi = \ln l \quad (\alpha = 1,\ldots,m) \tag{24}$$

Here $\pi_\alpha(\mathbf{x})$ denotes the α-th component of the stationary vector π of the Markov chain with transition matrix **Q** whose elements depend on the monomer mixture composition in microreactor **x** according to formula (8). To have the set of Eq. (24) closed it is necessary to determine the dependence of **x** on **X** in the thermodynamic equilibrium, i.e. to solve the problem of equilibrium partitioning of monomers between microreactors and their environment. This thermodynamic problem has been solved within the framework of the mean-field Flory approximation [48] for copolymerization of any number of monomers and solvents. The dependencies $x_\alpha = F_\alpha(\mathbf{X})(\alpha=1,\ldots,m)$ found there in combination with Eqs. (24) constitute a closed set of dynamic equations whose solution permits the determination of the evolution of the composition of macroradical $\mathbf{X}(l)$ with the growth of its length l, as well as the corresponding change in the monomer mixture composition in the microreactor.

In the model under consideration all macromolecules of fixed length l have the same composition $\mathbf{X}(l)$. However, owing to the substantial polydispersity of the products of free-radical copolymerization for length (even for polymer specimens obtained under low conversions), their composition distribution:

$$\langle \delta[\zeta - \mathbf{X}(l)] \rangle \equiv \frac{1}{\bar{l}} \int_1^\infty f(l)\,\delta[\zeta - \mathbf{X}(l)]\,dl \tag{25}$$

may turn out to be rather wide.

It is easy to notice a certain formal resemblance between this expression and the expression (11) for the composition inhomogeneity of the products of high-conversion copolymerization describable by the ideal model. In both expressions angular brackets denote the operation of averaging the bracketed quantity either over the conversion under which the polymer chain formed or over its length. This resemblance is also retained for the calculation of sequence distribution where only the quantities $\langle P\{U_k\}\rangle$ are obtainable experimentally. In both cases it implies the performance of corresponding averaging of instantaneous probabilities $P\{U_k\}$ in the mixture of fractions which are monodisperse in composition. Expressions for $P\{U_k\}$ within the framework of the model in hand will be identical to those occurring in the ideal model with the sole distinction that the components x_α of the monomer mixture composition vector involved in these expressions will be controlled by the macromolecule length l in a much different fashion than they were dependent on conversion in the case of the ideal model. The trajectory $\mathbf{x}(l)$ for the model of interest is prescribed not only by reactivities $\{r_{\alpha\beta}\}$ but also by thermodynamic parameters $\{\chi_{\alpha\beta}\}$ which characterize in the theory of solutions the pair interactions between monomers and monomeric units.

5.2 Living Anionic Copolymerization

The fundamental distinction between this process and free-radical copolymerization becomes apparent even in the framework of the ideal kinetic model. This is due to the perceptible change in monomer composition which occurs in the reaction system in the course of a polymer chain propagation. This phenomenon inevitably results in intramolecular chemical inhomogeneity of the products of high-conversion "living" anionic copolymerization [49, 50]. It show up dramatically the more strongly pronounced is the distinction in reactivities of monomers. If this distinction is large enough the products formed for such a process at complete conversion (p=1) are so-called tapered copolymers [51, 52]. Their macromolecules differ from those of ideal block copolymers in two features. Firstly, the border between blocks is more or less fuzzy and, secondly, the very blocks can comprise a number of incorporated units of the other monomer.

An exhaustive statistical description of "living" copolymers is provided in the literature [25]. There, proceeding from kinetic equations of the ideal model, the type of stochastic process which describes the probability measure on the set of macromolecules has been rigorously established. To the state $S_\alpha(\tau)$ of this process monomeric unit $\overline{\mathrm{M}}_\alpha$ corresponds formed at the instant τ by addition of monomer M_α to the macroradical. To the statistical ensemble of macromolecules marked by the label τ there corresponds a Markovian stochastic process with discrete time but with the set of transient states $S_\alpha(\tau)$ constituting continuum. Here the fundamental distinction from the Markov chain (where the number of states is discrete) is quite evident. The role of the probability transition matrix in characterizing this chain is now played by the integral operator kernel:

$$K_{\alpha\beta}(\tau,\eta)=\exp\left\{-\int_\tau^\eta \sigma_\alpha(\xi)d\xi\right\}k_{\alpha\beta}M_\beta(\eta),\quad (\alpha,\beta=1,\dots,m) \tag{26}$$

which has a sense of the density of the probability of transition from transient state $S_\alpha(\tau)$ into the state $S_\beta(\eta)$.

The kernel (26) in combination with the probability of the transition from the state $S_\alpha(\tau)$ to the absorbing state:

$$T_{\alpha 0}(\tau,t)=\exp\left\{-\int_{\tau}^{t}\sigma_\alpha(\xi)d\xi\right\},\text{ where }\ \sigma_\alpha(\xi)\equiv\sum_\beta k_{\alpha\beta}M_\beta(\xi) \tag{27}$$

completely defines the Markovian stochastic process in hand.

The kernel (26) and the absorbing probability (27) are controlled by the rate constants of the elementary reactions of chain propagation $k_{\alpha\beta}$ and monomer concentrations $M_\alpha(\tau)$ at the moment τ. These latter are obtainable by solving the set of kinetic equations describing in terms of the ideal kinetic model the alteration with time of concentrations of monomers M_α and reactive centers R_α.

$$\frac{dM_\alpha}{dt}=-\sum_\beta k_{\beta\alpha}R_\beta M_\alpha,\quad \frac{dR_\alpha}{dt}=\sum_\beta k_{\beta\alpha}R_\beta M_\alpha-\sum_\beta k_{\alpha\beta}R_\alpha M_\beta \tag{28}$$

Knowing the functions (26) and(27) it is possible by means of the formalism of the theory of Markovian processes [53] to find any statistical characteristic in an ensemble of macromolecules with labeled units. A subsequent label erasing procedure is carried out by integration of the obtained expressions over time of the formation of monomeric units. Examples of the application of this algorithm are reported elsewhere [25].

6 Step Polymerization

In step polymerization processes the functional groups act as reactive centers. Here the elementary reaction of chain propagation during polycondensation is represented by the interaction of two such groups, A_i and A_j, pertaining to different molecules, which leads to the formation of chemical bond Q_{ij}. As a result of this elementary reaction the combination of a pair of molecules takes place often accompanied by the formation of a molecule of a low molecular weight by-product, z_{ij}. Under certain conditions a back reaction is quite conceivable accompanied by the cleavage of a chemical bond with splitting of the polymer molecule. When the rate of this degradation reaction W_d is small compared with that of the reaction of condensation W the polycondensation process is found to proceed practically irreversibly. Under regime of reversible polycondensation, when the rates W_d and W are comparable, it is essential describing the kinetics of such a process to make allowance for the destruction of polymer molecules. If polymer synthesis is conducted under invariable conditions after a lapse of time the reaction system attains chemical equilibrium characterized by a particular distribution of the polycondensation products for size, composition and functionality. The calculation of this equilibrium SCFD may be accomplished by means of a thermodynamic method which is a far more simple task in comparison to the calculation of such a distribution in the case of irreversible polycondensation. Under equilibrium the rates of condensation and destruction have the same value *W*, controlled by the initial concentrations of the monomers and the temperature. If the by-product is withdrawn from the reaction zone at a rate W_{out} which is small relative to *W* at every instant the system will be practically in equilibrium. The parameters of the equilibrium SCFD will be governed by the instantaneous value of the by-product concentration in the reaction zone. Such a process of reversible polycondensation proceeding at $W_{out} << W$ will apparently be equilibrium.

The topological structure of condensation polymers is predetermined by the functionality of the initial monomers. If all of them are bifunctional then linear polymers are known to form. Branched and crosslinked molecules are prepared only when at least one of the monomers involves three or more functional groups.

When making theoretical considerations of polycondensation processes it is necessary to distinguish chemically identical functional groups if they differ in reactivity. Examples are primary and secondary hydroxyls in a molecule of glycerine, $SA^2{}_1A_2$, which belong to kinetically distinct types A_1 and A_2.

Monomers employed in a polycondensation process in respect to its kinetics can be subdivided into two types. To the first of them belong monomers in which the reactivity of any functional group does not depend on whether or not the remaining groups of the monomer have reacted. Most aliphatic monomers meet this condition with the accuracy needed for practical purposes. On the other hand, aromatic monomers more often have dependent functional groups and, thus, pertain to the second type. Obviously, when selecting a kinetic model for the description of polycondensation of such monomers, the necessity arises to take account of the "substitution effects" whereas the polycondensation of the majority of monomers of the first type can be fairly described by the ideal kinetic model. The latter, due to its simplicity and experimental verification for many systems, is currently the most commonly accepted in macromolecular chemistry of polycondensation processes.

6.1 Linear Polycondensation

Alongside the radical distinction of the mechanism of this process from that of chain polymerization, linear polycondensation features a number of specific peculiarities. So, for instance, the theory of copolycondensation does not deal with the problem of the calculation of a copolymer composition which normally coincides with the initial monomer mixture composition. Conversely, unlike chain polymerization, of particular importance for the products of polycondensation processes with the participation of asymmetric monomers is structural isomerism, so that the fractions of the "head-to-head"and "head-to-tail" patterns of arrangement prove generally to be comparable. Consequently, a necessity arises to solve the specific problems associated with the calculation of the microstructure parameters of polycondensation copolymers with allowance for various types of isomerism of their molecules. When considering polycondensation systems of dissimilar kinds questions need to be answered concerning the number of such parameters and the ways of selection best suited for the treatment of the experimental data found, for example, by an NMR spectroscopy technique. Once these structural parameters have been chosen the theory comes up against the problem of establishing their dependence on the stoichiometric, kinetic and thermodynamic parameters that characterize a reaction system. Substantial progress has been made in the solution of this problem [12, 54]. Further refinement of the theory of linear polycondensation is connected with the calculation of other statistical characteristics of the products of this process which describe their composition inhomogeneity and the sequence distribution in their macromolecules.

6.1.1
Monomers with Kinetically Independent Functional Groups

As the result of theoretical consideration of polycondensation of an arbitrary mixture of such monomers it was proved [55, 56] that the alternation of monomeric units along polymer molecules obey the Markovian statistics. If all initial monomers are symmetric, i.e. they resemble $A_\alpha S_\alpha A_\alpha$, units $S_\alpha(\alpha=1,\ldots,m)$ will correspond to the transient states of the Markov chain. The probability $\nu_{\alpha\beta}$ of transition from state S_α to S_β is the ratio $Q_{\alpha\beta}/\nu_\alpha$ of two quantities $Q_{\alpha\beta}$ and ν_α which represent, respectively, the number of dyads $(S_\alpha S_\beta)$ and monads (S_α) per one monomeric unit. Clearly, $Q_{\alpha\beta}$ is merely a ratio of the concentration of chemical bonds of the $\alpha\beta$-th type, formed as a result of the reaction between group A_α and A_β, to the overall concentration of monomeric units. The probability $\nu_{\alpha 0}$ of a transition from the transient state S_α to an absorbing state S_0 equals $1-p_\alpha$ where p_α represents the conversion of groups A_α.

Somewhat more complicated is the Markov chain describing the products of polycondensation with participation of asymmetric monomers. Any of them, $A_i S_\alpha A_j$, comprises a "tail-to-head" oriented monomeric unit S_α. It has been demonstrated [55, 56] that the description of molecules of polycondensation copolymers can be performed using the Markov chain whose transient states correspond to the oriented units. A transient state of this chain $\{ij\}$ corresponds to a monomeric unit at the left and right edge of which the groups A_i and A_j are positioned, respectively. A state $\{ji\}$ corresponds here to the same unit but is oriented in the opposite direction. However, a drawback of this Markov chain worthy of mention is the excessive number of its states.

It is possible, however, to eliminate this drawback [56] by enlarging the above Markov chain through a combination of several of its states into a single one. Such an enlargement is attainable in two ways. Following the first of them it is necessary as a transient state (j) of the enlarged chain to choose the sum of states $\{1j\}+\{2j\}+\ldots+\{mj\}$, whereas the second way suggests that as such a state (i) the combination of states $\{i1\}+\{i2\}+\ldots+\{im\}$ acts. The enlarged Markov chains obtained in these two ways discriminate between "heads" and "tails" either only on right edges or only on left edges of monomeric units, respectively. The probabilities of the transitions between transient states of the first of these Markov chains are the elements of a matrix representing the product of two square matrices **p** and **H**. The first of them has elements $p_{ij}=Q_{ij}(1+\delta_{ij})/\mu^0_i$, where μ^0_i is the number of functional groups A_i per one monomeric unit in molecules of the initial monomers, while δ_{ij} denotes the Kronecker delta-symbol. The element H_{ij} of the matrix **H** is distinct from zero only provided that the initial mixture comprises monomer $A_iS_\alpha A_j$. In this case $H_{ij}=\nu_\alpha(1+\delta_{ij})$. The probability of the transition ν_{j0} from a transient state (j) to an absorbing one for the "right edge" Markov chain under consideration is equal to $1-p_j$. An analogous "left edge" Markov chain has also been described [12].

It is common knowledge that in order to synthesize high molecular weight products by polycondensation it is necessary to choose fractions of the initial monomers so as to ensure that the deviations from their stoichiometric condition, providing a complete conversion $p_i=1$ of all functional groups A_i, are minimal [57]. Under this condition for solving the problem of the sequence distribution it is possible to neglect the finiteness of macromolecules and to calculate the fractions of dyads, triads and so on resorting to the formalism of ergodic Markov chains rather than to the formalism of absorbing ones.

Noteworthy that all the above formulated results can be applied to calculate the statistical characteristics of the products of polycondensation of an arbitrary mixture of monomers with kinetically independent groups under any regime of this process. To determine the values of the elements of the probability transition matrix of corresponding Markov chains it will suffice to calculate only the concentrations Q_{ij} of chemical bonds (ij) at different conversions of functional groups. In the case of equilibrium polycondensation the concentrations Q_{ij} are controlled by the thermodynamic parameters, whereas under the nonequilibrium regime of this process they depend on kinetic parameters.

6.1.2
Monomers with Dependent Groups

When monomers with dependent groups are involved in a polycondensation, the sequence distribution in the macromolecules can differ under equilibrium and nonequilibrium regimes of the process performance. This important peculiarity, due to the violation in these nonideal systems of the Flory principle, is absent in polymers which are synthesized under the conditions of the ideal polycondensation model. Just this circumstance deems it necessary for a separate theoretical consideration of equilibrium and nonequilibrium polycondensation.

Such a consideration demonstrated [56] that the sequence distribution in products of arbitrary equilibrium copolycondensation can always be described by some Markov chain with the elements of the transition probability matrix expressible through thermodynamic and stoichiometric parameters of a system [58]. Yet under a nonequilibrium regime of the performance of this process the above assertion holds for quasiideal systems only [59]. Among these latter there are such nonideal systems, where every unit of a monomer with dependent groups in any polymer molecule is connected by chemical bonds exclusively with units of the monomers containing independent groups.

The simplest system of those which do not meet this condition is the homopolycondensation of monomer SA^2 whose groups are dependent. As a theoretical analysis of this process proceeding under the equilibrium regime has demonstrated [2, 60], the molecular weight distribution $f(l)$ of the polymers formed is described as in the case of the ideal model by the Flory formula. This exponential distribution can be derived using the statistical method when polymer molecules are envisaged as realizations of an absorbing Markov chain with the only transient state. A completely different type of situation occurs under the irreversible regime of polycondensation when the distribution $f(l)$ calculated within the framework of the same model differs materially from the exponential one [2]. It follows from the numerical calculations [61] at certain values of kinetic parameters that this distribution may be bimodal and even oscillating in shape. In terms of the statistical method it was rigorously proved [61] that in this case a random process of conventional movement along a macromolecule may not be described by a Markov chain of any finite order. Theoretical examination of the effect of the dependence of functional groups in monomer molecules on the character of the sequence distribution in the products of irreversible copolycondensation has been exhaustively carried out for the interbipolycondensation of two comonomers $A_1R_1A_1$ and $A_2R_2A_2$ with intercomponent BIB. The simplest characteristic defining the structure of chains formed in such a system of binary copolymers $\sim IR_1IR_1IR_2IR_2\sim$ is a microheterogeneity coefficient K_M which depends only on fractions of dyads of elementary units (R_1I) and (R_2I). Kinetic calculation of this structural parameter showed [62] that the character of a chain microstructure is predetermined to a great extent by the dependence of the reactivities of functional groups in an intercomponent. If they are independent according to the theory $K_M=1$ and the sequence distribution at any conversion, the stoichiometry of the initial mixture of monomers and their reactivity is always random. A general correlation has been established between the fashion of the change of the reactivity of intercomponent functional groups in the course of their reaction with comonomers and the manner of the sequence distribution in a copolymer [62]. If entering into the reaction of the first group of an intercomponent is accompanied by a decrease in the reactivity of its second group the theory predicts the tendency of chain units to alternation ($K_M>1$). In the opposite case, under the growth of the reactivity of the second group, a tendency to block formation is expected to arise ($K_M<1$). These theoretical regularities have been verified for the products of acceptor-catalytic interbipolycondensation by comparing the values of the quantity K_M found from the analysis of spectroscopic data and those calculated theoretically [54, 63].

For the interbipolycondensation the condition of quasiideality is the independence of the functional groups either in the intercomponent or in both comonomers. In the first case the sequence distribution in macromolecules will be described by the Bernoulli statistics [64] whereas, in the second case, the distribution will be characterized by a Markov chain. The latter result, as well as the parameters of the above mentioned chain, were firstly obtained within the framework of the simplified kinetic model [64] and later for its complete version [59]. If all three monomers involved in interbipolycondensation have dependent groups then, under a nonequilibrium regime, non-Markovian copolymers are known to form.

6.2 Branched Polycondensation

6.2.1 *Ideal Model*

Let us address first the processes of irreversible polycondensation of an arbitrary mixture of monomers. The functional groups $A_1,\ldots,A_i,\ldots,A_m$ act here as kinetically independent elements, and the scheme of the elementary reactions of condensation between them

$$A_i + A_j \xrightarrow{k_{ij}} Q_{ij} + z_{ij} \qquad \left(1 \le i, j \le m\right) \tag{29}$$

is characterized by $m(m+1)$ kinetic parameters $\{k_{ij}\}$.

The rate constants in the reactions (29) may be conveniently envisaged as elements of symmetric matrix **k**. In order to calculate the statistical characteristics of a particular polycondensation process along with matrix **k** parameters should be specified which characterize the functionality of monomers and their stoichiometry. To this end it is necessary to indicate the matrix **f** whose element $f_{i\alpha}$ equals the number of groups A_i in an α-th type monomer as well as the vector **v** with components $v_1,\ldots v_\alpha,\ldots,v_n$ which are equal to molar fractions of monomers $M_1,\ldots,M_\alpha,\ldots,M_n$ in the initial mixture. The general theory of polycondensation described by the ideal model was developed more than twenty years ago [2]. Below the key results of this theory are presented.

6.2.1.1 Kinetic Method

The composition and the functionality of a polymer molecule are defined by the numbers $l_\alpha(\alpha=1,\ldots,n)$ and $a_i(i=1,\ldots,m)$ of monomeric units and functional groups of different types involved in the molecule. Considering these numbers as components of vectors **l** and **a** in order to denote the polymer molecule it is convenient to introduce symbol {**l**;**a**}. In terms of the latter, taking into account the scheme of the elementary reactions in Eq. (29), it is a simple task to describe the scheme of the reactions between molecules:

$$\{\mathbf{l}';\mathbf{a}'\}+\{\mathbf{l}'';\mathbf{a}''\}\rightarrow\{\mathbf{l}'+\mathbf{l}'';\mathbf{a}'+\mathbf{a}''-\mathbf{e}_i-\mathbf{e}_j\}+z_{ij} \tag{30}$$

Here all components of the vector $\mathbf{e}_i$, except the i-th one which is unity, are equal to zero. Because the molecular reaction (30) is induced by chemical interaction of groups A_i and A_j its rate constant apparently equals $k_{ij}(a_i'a_j''+a_j'a_i'')$. It is possible to write down an infinite set of kinetic equations corresponding to the scheme (30) for the concentrations C(l;a;t) of molecules of different composition and functionality which are present in the reaction system at the moment t. To solve these equations it is convenient to go over to the equivalent equation:

$$\frac{\partial g}{\partial t}=M\sum_{i=1}^{m}\sum_{j=1}^{m}k_{ij}\left(-\mu_j x_i\frac{\partial g}{\partial x_i}+\frac{1}{2}\frac{\partial g}{\partial x_i}\frac{\partial g}{\partial x_j}\right) \tag{31}$$

$$g(\mathbf{s};\mathbf{x};t)\equiv\sum_{\mathbf{a}}\sum_{\mathbf{l}}C(\mathbf{l};\mathbf{a};t)\prod_{\alpha=1}^{n}s_\alpha^{l_\alpha}\prod_{i=1}^{m}x_i^{a_i} \tag{32}$$

of the distribution of reagents for size, composition and functionality (SCFD). Their concentration entering into the expression (32) is made dimensionless by the division by the overall concentration M of monomeric units. The change with time of the average number μ_i of functional groups A_i per one monomeric unit can be found from the solution of the following equations:

$$\frac{d\mu_i}{dt}=-M\mu_i\sum_{j=1}^{m}k_{ij}\mu_j,\quad \mu_i(0)\equiv\mu_i^0=\sum_{\alpha=1}^{n}\mathrm{f}_{i\alpha}\nu_\alpha \tag{33}$$

Exact solution of the partial differential equation (31) under the initial condition

$$g(\mathbf{s};\mathbf{x};0)\equiv h(\mathbf{s};\mathbf{x})=\sum_{\mathbf{a}}\sum_{\mathbf{l}}C(\mathbf{l};\mathbf{a};0)\prod_{\alpha=1}^{m}s_\alpha^{l_\alpha}\prod_{i=1}^{m}x_i^{a_i} \tag{34}$$

obtained by the method of characteristics will read

$$g=h(\mathbf{s};\xi)-\frac{1}{2}\sum_{i=1}^{m}\sum_{j=1}^{m}L_{ij}h_i(\mathbf{s};\xi)h_j(\mathbf{s};\xi) \tag{35}$$

where the following designations are used

$$L_{ij}=\frac{M}{\mu_i^0\mu_j^0}\int_0^t k_{ij}\mu_i\mu_j dt',\qquad h_i(\mathbf{s};\xi)=\frac{\partial h}{\partial x_i}\Big|_{\mathbf{x}=\xi} \tag{36}$$

Expressions (35), (36) in conjunction with characteristics

$$\left(1-p_i\right)x_i=\xi_i-\sum_j L_{ij}h_j\text{, where }\quad p_i\equiv 1-\frac{\mu_i}{\mu_i^0}=\sum_j L_{ij}\mu_j^0 \tag{37}$$

which establish the correlation between vectors ζ and $\mathbf{x}$, define gf (32). Another two generating functions are expressed through it

$$G(\mathbf{s})\equiv g(\mathbf{s};1)\qquad G_W\left(\mathbf{s}\right)=\sum_\alpha s_\alpha\frac{\partial G}{\partial s_\alpha}=\sum_{\mathbf{l}} f_W\left(\mathbf{l}\right)\prod_\alpha s_\alpha^{l_\alpha} \tag{38}$$

The first of them occurs in thermodynamic theory of solutions and blends of heteropolymers [3] whereas the second one is the gf of the weight SCD $f_W(\mathbf{l})$. It can readily be determined from the simple expression:

$$G_W\left(\mathbf{s}\right)=h\left(\mathbf{s};\xi\right)=\sum_\alpha \nu_\alpha s_\alpha\prod_i \xi_i^{\mathrm{f}_{i\alpha}} \tag{39}$$

where the dependence of the components of the vector ζ on s can be determined by solving the following set of algebraic equations

$$\xi_i=1-p_i+\sum_j\sum_\alpha p_{ij}\kappa_{j\alpha}s_\alpha\prod_k \xi_k^{\mathrm{f}_{k\alpha}-\delta_{kj}} \tag{40}$$

Here δ_{kj} is the Kronecker delta-symbol. Apart from components $\mathrm{f}_{k\alpha}$ of the matrix of functionalities $\mathbf{f}$ and conversion p_i of the groups A_i (37) to the right-hand part of Eqs. (40), there also enter the elements p_{ij} and $\kappa_{j\alpha}$ of the matrix of bonds $\boldsymbol{p}$ and the stoichiometric matrix κ, respectively

$$p_{ij}=\frac{Q_{ij}}{\mu_i^0}\left(1+\delta_{ij}\right)=L_{ij}\mu_j^0\qquad \kappa_{j\alpha}=\frac{\mathrm{f}_{j\alpha}\nu_\alpha}{\mu_j^0} \tag{41}$$

To establish the dependence of quantities p_{ij} on time it is necessary to determine the dimensionless concentrations Q_{ij} bonds per monomeric unit. It is sufficient for this purpose to solve Eqs. (33) and to calculate the integral (36). The element $\kappa_{j\alpha}$ of the stoichiometric matrix, according to (41), represents the fraction of all groups A_j in the initial mixture which belong to monomer M_α.

A straightforward scenario for determining the gf of the weight SCD (38) of polycondensation polymers consists in finding the solution $\zeta(\mathbf{s})$ of the algebraic equations (40), its subsequent substitution into formula (39) and the expansion of the expression obtained into a power series of the vector s components. However, another way, which is based on the special relationships for the derivatives of the implicit functions, was found to be more convenient [65]. This operation works even when it is impossible to find the explicit solution to Eqs. (40). Equations (39) and (40) permit one, in a rather simple manner, to derive the formulas for the statistical moments of the SCD avoiding the highly tedious procedure of determination of this very distribution. Hence, for instance, the weight average degree of polymerization can be calculated using the following equation:

$$P_W \equiv \sum_l l f_W(l) = \sum_\alpha \frac{\partial G_W}{\partial s_\alpha}\Big|_{\mathbf{s}=1} = 1 + \sum_i \sum_j \mu_i^0 \Delta_{ij} p_j \tag{42}$$

where Δ_{ij} are elements of matrix Δ

$$\Delta = (\mathbf{E} - \mathbf{pH})^{-1}, \quad \mathbf{H} = \kappa \mathbf{f}^T - \mathbf{E} \tag{43}$$

and $\mathbf{f}^T$ is the matrix transposed to **f**. The gel-point condition $P_W=\infty$, as determined from expressions (42) and(43), is the vanishing of the matrix (**E-pH**) determinant.

6.2.1.2 Statistical Method

Equations (39) and (40) permit a simple probabilistic interpretation in terms of the theory of branching processes which describe the populations of independently reproducing particles. Under such a consideration to each of them a particular fragment of the polymer molecule corresponds. Noteworthy, this correspondence may differ for various modifications of the statistical method [66]. For the modification to be considered below there are two classes of reproducing particles. The first of them corresponds to functional groups A_i, including the reacted ones, while the second corresponds to monomeric units S_α labeled by "color" j, which indicates the type of the reacted group A_j in this unit. A population of such a branching process develops in the following way. Firstly, a particle ancestor of the α-th type gives birth in the zero-th generation to the descendants of the first class among which $f_{i\alpha}$ particles pertain to the i-th type. The probability generating function in this generation is as follows:

$$F^{(0)}(\mathbf{y}) = \sum_\alpha \nu_\alpha s_\alpha \prod_i y_i^{f_{i\alpha}} \tag{44}$$

Each particle (i) of the zero-th generation can either die with the probability $1-p_i$ or give birth to a particle of the second class ($j\alpha$) with probability $p_{ij}\kappa_{j\alpha}$. Consequently, the gf of a particle belonging to the first generation is defined as:

$$\Phi_i(\mathbf{s}^c) = 1 - p_i + \sum_j \sum_\alpha p_{ij} \kappa_{j\alpha} s^c_{(j\alpha)} \tag{45}$$

where $S^c_{(j\alpha)}$ is the marker variable for the random appearance of particle ($j\alpha$) associated with the colored monomeric unit S^j_α. Evidently, the descendants of such a particle of the second generation will differ from those produced by the particle ancestor by the absence of one particle (j), so that the gf of the second class particle is represented by the formula:

$$F_{(j\alpha)}(\mathbf{y}) = \prod_k y_k^{f_{k\alpha} - \delta_{kj}} \tag{46}$$

For subsequent evolution of the population the events mentioned are recursively repeated. According to the definition of the Galton-Watson branching process, the probabilities of the reproduction do not depend here on the number of generations, so that expressions (45) and (46) are valid for all generations beginning with the first one. In line with the general theory of such stochastic processes [16] the *gf* of the probability distribution of populations with a given set of numbers of particles is defined by the following formulas:

$$G_W^c\left(\mathbf{s}^c;\mathbf{y}\right)=F^{(0)}(\xi), \qquad \xi_i=y_i\Phi_i\left(\mathbf{u}^c\right), \qquad u_{(j\alpha)}^c=s_{(j\alpha)}^c F_{(j\alpha)}(\mathbf{y}) \tag{47}$$

which completely characterize the extended branching process. To eliminate excessive information contained in these formulas one should set y_i=1 and $s^c_{(j\alpha)}=s_\alpha$ should be set for all values of the indices, i.e. not to distinguish molecules by the number of groups and to "erase" colors which mark monomeric units. It can be easily verified that expressions (44)–(47) will go over to (39) and (40).

Depending on the values of the probability parameters of the branching process two paths of evolution for every population are possible: either it will degenerate in some generation or it will infinitely grow. The probability $\xi^*_1,...,\xi^*_m$ of degeneration of a population generated by particles (1),...(m) is the solution of the set of algebraic equations (40). Up to the gel-point $\xi^*_1=...=\xi^*_m=1$, whereas after the gel-point $\xi^*_i<1$ at all values of *i*. Knowing the components ξ^*_i of the vector ξ^* it is easy to calculate the fraction of all monomeric units involved in the sol-fraction ω_s as well as its composition $\mathbf{X}^s$

$$\omega_s=\sum_\alpha \nu_\alpha \prod_i \left(\xi_i^*\right)^{f_{i\alpha}} \qquad X_\alpha^s=\nu_\alpha \prod_i \left(\xi_i^*\right)^{f_{i\alpha}} / \omega_s \tag{48}$$

It is worth emphasizing that formulas (39) and (40) may be invoked to determine the statistical characteristics of the sol-fraction provided the probability parameters are replaced in these equations by their modified values

$$\hat{\nu}_\alpha=\nu_\alpha \prod_i \left(\xi_i^*\right)^{f_{i\alpha}} \qquad \hat{p}_{ij}\hat{\kappa}_{j\alpha}=p_{ij}\kappa_{j\alpha}\prod_k \left(\xi_k^*\right)^{f_{k\alpha}} / \xi_i^*\xi_j^* \tag{49}$$

The relationship established between stochastically branched molecules of Gordonian polycondensation polymers and the realizations of the corresponding branching process provides the possibility to find in parallel with the SCD the fraction of any isomer [67] and that of the arbitrary fragment of molecules that characterize the microstructure of these latter [68]. Central among the statistical characteristics of a gel is its cyclic rank, responsible for the elasticity of the polymer network. A formula has been derived [69] for this statistical parameter of the global topological structure of the molecular graph of a network formed during polycondensation of an arbitrary monomer mixture provided the ideal model is applicable.

6.2.1.3
Universality

The theoretical results reported in the foregoing referred to irreversible polycondensation because just for this regime the derivation of expressions (39), (40) has been demonstrated by means of kinetic method. Proceeding from thermodynamic considerations it was rigorously proved [70] that the products of equilibrium polycondensation are describable by these very expressions. A correlation has also been disclosed between probability parameters p_{ij} of the branching process and the equilibrium constants of elementary reactions (29). Since under two limiting kinetic regimes (for one of which the destruction rate is negligible whereas for the other it attains maximum value) the polycondensation products are described by the same branching process, there are serious grounds to deem it to be universal for any synthesis regime. The probability parameters p_{ij} in the general case are related to the concentrations Q_{ij} of chemical bonds by the relationship (41), while values Q_{ij} can be calculated by solving the set of material balance equations [2] corresponding to the elementary reaction (29) with allowance for its reversibility. Hence, the quantitative theory of polycondensation described by the ideal kinetic model may be envisaged today as being complete.

6.2.2
First-Shell Substitution Effect Model

The reactivity of a functional group can be dependent not only on its type but also on the microstructure of the molecule fragment to which this group is attached. If such a fragment is a monomeric unit with adjacent chemical bonds then the functional group reactivity will be governed solely by the number of such bonds, their types and their mutual arrangement in the unit comprising the reacting group. Within the framework of this approximation account is taken only of the "first-shell" substitution effect (FSSE). When the presence of a reacted group also affects the reactivity of groups belonging to the next monomeric unit then the term "second-shell" substitution effect is used, and so on. Here only FSSE will be considered.

The key peculiarity of nonideal models is the fact that unlike for the case of the ideal model they do not exhibit universality. In other words the products of equilibrium and irreversible polycondensation are characterized in the framework of the FSSE model by different branching processes. This assertion will be exemplified below by the homopolycondensation of a monomer SA^f containing f identical functional groups A. Under theoretical consideration of this system it is convenient to distinguish monomeric units S by their kind depending on the number i of reacted groups attached to them. This means that a unit belongs to the kind i if it has i adjacent chemical bonds connecting this unit with the neighboring ones.

6.2.2.1 Equilibrium Polycondensation

Here the FSSE model is characterized by f thermodynamic parameters $k_0,\ldots,k_1,\ldots,k_{f-1}$ where k_i is the equilibrium constant of the elementary reaction between functional groups attached to the i-th type unit and to the monomer. It has been shown [71, 72] that it is possible to calculate the statistical characteristics of the polycondensation products via the Galton-Watson branching process with one type of reproducing particles corresponding to either chemical bonds or reacted functional groups. Generating functions of the distribution of probabilities of the number of descendants in the zero-th and all the subsequent generations

$$F^{(0)}(s)=\sum_{i=0}^{f}\lambda_i s^i, \qquad F(s)=\frac{dF^{(0)}(s)}{ds}\left[\frac{dF^{(0)}(s)}{ds}\right]_{s=1}^{-1}=\sum_{i=0}^{f}d_i s^{i-1} \tag{50}$$

are defined exclusively by the number, λ_i, and weight, $d_i=i\lambda_i/p\mathrm{f}$, distribution of monomeric units for their kind i. Substitution of the expressions in Eq. (50) into formulas of the Galton-Watson process [16] for gf of the distribution of number of all particles constituting the population

$$G_{pop}(s)=sF^{(0)}(u) \qquad u=sF(u) \tag{51}$$

leads to the expression for generating function $G_W(s)$ of weight MWD of the products of equilibrium polycondensation.

$$G_W(s)=s\sum_{i=0}^{f}\lambda_i u^i \qquad u=s\sum_{i=1}^{f}d_i u^{i-1} \tag{52}$$

Fractions λ_i of the i-th type of monomeric units can be calculated starting from the following expressions [73]:

$$\begin{gathered}\lambda_i=\frac{\Delta_i}{\Delta}, \quad \Delta=\sum_{i=0}^{f}\Delta_i, \quad \Delta_i=\frac{\mathrm{f}!}{i!(\mathrm{f}-i)!}\Phi(i)\omega^i \\ \Phi(0)=\Phi(1)=1, \quad \Phi(i)=\prod_{j=1}^{i-1}\kappa_j, \quad (2\le i\le \mathrm{f}), \quad \kappa_j\equiv\frac{k_j}{k_0}\end{gathered} \tag{53}$$

The auxiliary variable ω is uniquely related to the conversion:

$$p=\frac{1}{\mathrm{f}}\sum_{i=1}^{f}i\lambda_i=\frac{\omega}{\mathrm{f}}\frac{d}{d\omega}\ln\Delta \tag{54}$$

so that in order to calculate all the statistical characteristics of a polymer chemical structure at fixed values of p it will suffice to know the values of the independent thermodynamic parameters $\kappa_1,...\kappa_f$.

The weight fraction of sol can be calculated from the formulas:

$$\omega_s = \sum_{i=0}^{f} \lambda_i \left(u^*\right)^i \qquad u^* = \sum_{i=1}^{f} d_i \left(u^*\right)^{i-1} \tag{55}$$

where the probability u^* of degeneration of a population generated by one particle is defined as the smallest positive root of Eq. (55). The calculation of MWD and its statistical moments can be carried out, following Dobson and Gordon [74], in the same manner as normally done before gelation, by switching to modified generating functions:

$$\hat{F}^{(0)}(s) = F^{(0)}\left(u^* s\right) / \omega_s \qquad \hat{F}(s) = F\left(u^* s\right) / F\left(u^*\right) \tag{56}$$

A general theory of the equilibrium polycondensation of an arbitrary mixture of monomers, described by the FSSE model, has been developed [75]. Proceeding from rigorous thermodynamic considerations a branching process has been indicated which describes the chemical structure of condensation polymers and expressions have been derived which relate the probability parameters of this stochastic process to the thermodynamic parameters of the FSSE model.

6.2.2.2
Irreversible Polycondensation

In this case the role of kinetically independent elements in accordance with the extended Flory principle is performed by monomeric units of different kinds $S_0,...,S_f$, while here the rate constants of the elementary reactions of condensation $k_{ij}(i,j=0,..., f-1)$ between functional groups belonging to units S_i and S_j act as the parameters of the FSSE model. It has been suggested [76] to characterize every molecule by vector $\mathbf{l}$ with components $l_0,...,l_i,...,l_f$ equal to the numbers of units of types $S_0,...,S_i,...,S_f$ that this molecule is comprised of. Kinetic schemes of the reactions between monomeric units as well as between molecules may be written down as follows:

$$S_i + S_j \rightarrow S_{i+1} + S_{j+1}; \quad \left\{\mathbf{l}'\right\} + \left\{\mathbf{l}''\right\} \rightarrow \left\{\mathbf{l}' + \mathbf{l}'' - \mathbf{e}_i - \mathbf{e}_j + \mathbf{e}_{i+1} + \mathbf{e}_{j+1}\right\} \tag{57}$$

The rate constants of these reactions are equal, respectively, to $\tilde{k}_{ij}=k_{ij}(f-i)(f-j)$ and $\tilde{k}_{ij}\,(l'_i l''_j + l'_j l''_i)$. For the generating function

$$g(\mathbf{s};t) = \sum_{\mathbf{l}} C(\mathbf{l};t) \prod_{i=0}^{f} s_i^{l_i} \tag{58}$$

of dimensionless concentration $C(\mathbf{l},t)$ of molecules with a given value of vector $\mathbf{l}$ the following equation has been derived [61, 76]:

$$\frac{\partial g}{\partial t}=M\sum_{i=0}^{f}\sum_{j=0}^{f}\tilde{k}_{ij}\left\{-\lambda_i s_j\frac{\partial g}{\partial s_j}+\frac{1}{2}s_{i+1}s_{j+1}\frac{\partial g}{\partial s_i}\frac{\partial g}{\partial s_j}\right\} \tag{59}$$

The values of molar fractions λ_i of units S_i and complete conversion of functional groups can be obtained by solving the following equations:

$$\frac{d\lambda_i}{dt}=\varphi_{i-1}\lambda_{i-1}-\varphi_i\lambda_i,\ where\quad \varphi_i\equiv M\sum_i\tilde{k}_{ij}\lambda_j$$
$$\lambda_i(0)=\delta_{i0}\qquad p=\frac{1}{f}\sum_j j\lambda_j \tag{60}$$

As distinct from the ideal model of irreversible polycondensation the exact analytical solution of Eq. (59) under initial condition $g(\mathbf{s};0)=s_0$ is hardly attainable. However, differentiating it term by term with respect to the components of vector $\mathbf{s}$ and setting all of them equal to unity it is possible to derive a closed set of ordinary differential equations for the second-order statistical moments of the distribution $C(\mathbf{l},t)$ and to solve them numerically to calculate the value of conversion at the gel-point $p=p^*$ where these moments turn to infinity. Such a calculation has been performed [77] for different sets of kinetic parameters $\{k_{ij}\}$. An alternative way of finding p^*, based on the analysis of a solution to Eq. (59) within the context of the theory of bifurcation, has been advanced [61]. The accuracy of the calculation of the gel-point has been ensured by a comparison of p^* values obtained in the two fashions.

A major problem which has long been a challenge for quantitative theory of polycondensation is whether the statistical method is applicable or not for the description of the chemical structure of the products of irreversible polycondensation of a monomer SA^f characterized by the FSSE model. First, the Galton-Watson branching process with generating functions (50) was believed [71] capable of providing a rigorous statistical description of these products as it is the case for equilibrium polycondensation. The erroneous character of this claim was disclosed more than twenty years ago [2]. Nevertheless the question concerning the existence of some alternative branching process which would be able to offer such a description has been an open question up to now. Today it is possible to rigorously substantiate the existence of such a process. To do this results presented in paper [61] should be rewritten as:

$$G_W(s)=s\mathcal{F}^{(0)}[\mathbf{u}]\qquad u_i(\tau)=s\mathcal{F}_i(\tau;[\mathbf{u}])\qquad (i=0,\ldots,f-1) \tag{61}$$

where the following designations are employed:

$$\mathcal{F}^{(0)}[\mathbf{u}]=\sum_{i=0}^{f}\int\ldots\int P_i(t;\{\tau_j\})\prod_{j=0}^{i-1}u_j(\tau_j)d\tau_j \tag{62}$$

$$\mathcal{F}_i(\tau;[\mathbf{u}]) = \frac{\delta \mathcal{F}^{(0)}[\mathbf{u}]}{\delta u_i(\tau)} \left[\frac{\delta \mathcal{F}^{(0)}[\mathbf{u}]}{\delta u_i(\tau)} \right]^{-1}_{\mathbf{u}(\tau)=1} \tag{63}$$

$$P_i(t;\{\tau_j\}) = \exp[-I_i(t)] \prod_{j=0}^{i-1} v_j(\tau_j)\varphi_j(\tau_j) \tag{64}$$

$$v_i(\tau) = \exp[I_{i+1}(\tau) - I_i(\tau)] \qquad I_i(t) = \int_0^t \varphi_i(\tau) d\tau \tag{65}$$

It is easy to see that the formulas (61) are similar to those (52) with the essential distinction that instead of generating functions (*gf*) $F^{(0)}(u)$ and $F(u)$, the generating functionals (*GF*) $\mathcal{F}^{(0)}[\mathbf{u}]$ (62) and $\mathcal{F}_i(\tau;[\mathbf{u}])$ (63) appear. To find the *gf* of the weight MWD it is necessary according to (61) to solve a set of integral equations for functions $u_i(\tau)$ and substitute the solution found into the expression for $\mathcal{F}^{(0)}[\mathbf{u}]$.

Relationships (61)–(63) admit simple probabilistic interpretation in terms of the branching process. To the reproducing particles of this process the reacted functional groups correspond distinguished by "color" i and label τ. Integer i characterizes the type S_i of monomeric unit to which a given group was attached at the moment τ of its formation.

Function $P_i(t;\{\tau_j\})$ (64) represents the density of probability for the particle ancestor at the moment t to give birth to i descendants (whose colors are $0,1,...,i-1$) at the instants $\tau_0<\tau_1<...<\tau_{i-1}$, respectively. In order to calculate at a given conversion the values of these functions it is sufficient to solve Eqs. (60) and make recourse to the formulas (65). Expression (63) relating *GF* $\mathcal{F}_i$ to $\mathcal{F}^{(0)}$ resembles expression (50) which relates *gf* F to $F^{(0)}$. The distinction consists the fact that the role of the ordinary derivative (59) is now played by functional (variational) derivatives (63).

Once the particular branching process that specifies the probability measure on the set of macromolecules of a polymer specimen has been identified, the statistical method provides the possibility to determine any statistical characteristic of the chemical structure of this specimen. In particular, the dependence of the weight fraction of a sol on conversion can be calculated by formulas [extending those (55)] which are obtainable from (61) provided the value of dummy variable s is put unity:

$$\omega_s = \mathcal{F}^{(0)}[u^*] \qquad u_i^*(\tau) = \mathcal{F}_i(\tau;[u^*]) \qquad (i=0,...,f-1) \tag{66}$$

To determine ω_s one should solve the set of f integral equations for probabilities of degeneration $u^*_0(\tau),...,u^*_{f-1}(\tau)$ and substitute these functions into functional $\mathcal{F}^{(0)}[\mathbf{u}]$ (Eq. 62). Numerical solution of these equations by means of the iteration method presents no difficulties since the integral operator is a contrac-

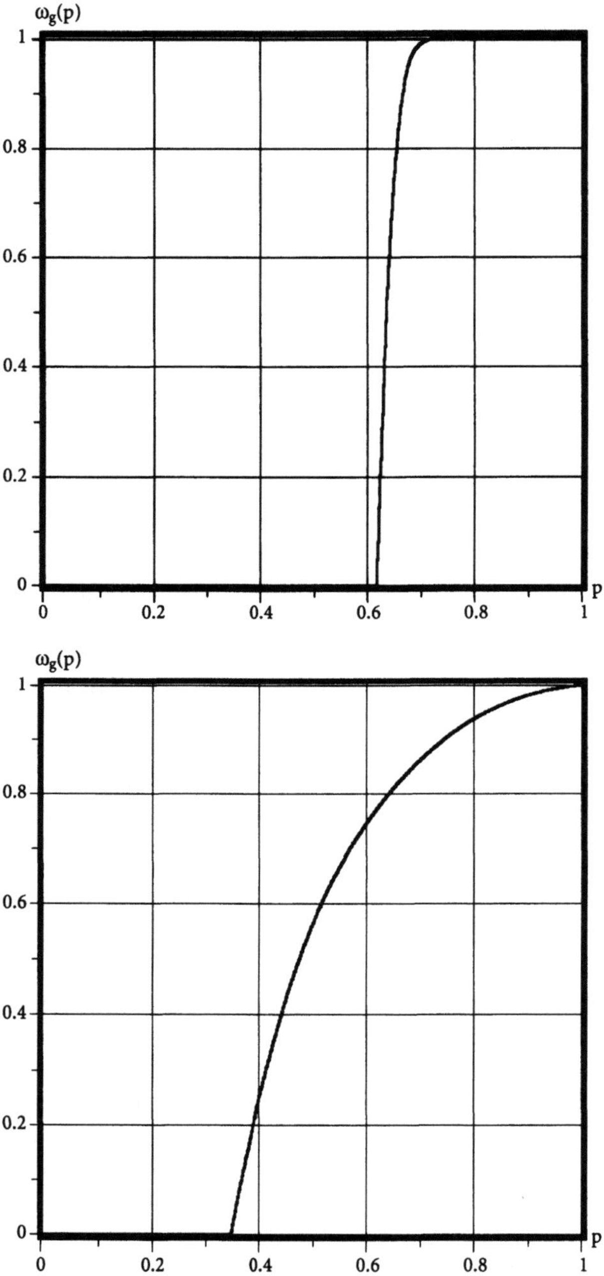

Fig. 1. The dependence of weight fraction of gel on conversion of functional groups A under irreversible polycondensation of monomer SA^3 described by the simplified FSSE model with kinetic parameters $\kappa_1 \equiv k_1/k_0$ and $\kappa_2 \equiv k_2/k_0$. The curves are depicted proceeding from the results of calculations at values of these parameters equal to $\kappa_1=1$, $\kappa_2=0.1$ (**a**), and $\kappa_1=1$, $\kappa_2=10$ (**b**)

tion one on the set of positive smooth vector-functions $\mathbf{u}(\tau)$. This ensures the convergence of the iterations to the sought for solution $\mathbf{u}^*(\tau)$. It is especially easy to find it in the case of the simplified FSSE model when the rate constants of elementary condensation reactions are factorizable $k_{ij}=k_ik_j$. Here the set of f equations for probabilities of degeneration is reduced to a single integral equation. Examples of the calculation of $\omega_g=1-\omega_s$ in the framework of such a simplified model at several values of kinetic parameters are depicted in Fig. 1. The calculation of other statistical characteristics of sol and gel can be found elsewhere [78].

The results reported above have been extended to the general case of irreversible polycondensation of an arbitrary mixture of monomers (characterized by arbitrary matrix of functionalities **f** and the composition vector **v**) under the conditions of the applicability of the FSSE model [26].

Acknowledgements. The author is very much indebted to the editor, Professor Ulrich Suter, for critically reading the manuscript and providing a number of valuable comments. Most sincere thanks are also due to the author's Portuguese friends Dr Mario Rui Costa and Dr Rolando Dias as well as Filomena, Luis and Emilia from the University of Porto for creating a special atmosphere which helped to alleviate the difficult task of the completion of this manuscript. The manuscript would not have been possible without the technical help of the author's son Ilya and wife Natasha to whom he owes his most sincere gratitude for their never-ending patience.

References

1. Lowry GG (ed) (1989) Markov chains and Monte Carlo calculations in polymer science. Marcel Dekker Inc, New York
2. Kuchanov SI (1978) Methods of kinetic calculations in polymer chemistry. Khimia Publ, Moscow
3. Kuchanov SI, Panyukov SV (1996) Statistical thermodynamics of heteropolymers and their blends. In: Allen G (ed) Comprehensive polymer science, 2nd suppl. Pergamon Press, New York, chap 13
4. Kuchanov SI, Panyukov SV (1998) J Polym Sci 36:937
5. Glockner G (1991) Gradient HPLC of copolymers and chromatographic cross-fraction. Springer, Berlin Heidelberg New York
6. Cowie JMG (1994) Macromol Symp 78:15
7. Guillot J, Emelie B (1991) Makromol Chem Rapid Commun 12:117
8. Bovey FA (1972) High resolution NMR of macromolecules. Academic Press, New York
9. Tonelli AE (1989) NMR spectroscopy and polymer microstructure. VCH, Weinheim
10. Koenig JL (1980) Chemical microstructure of polymer chains. John Wiley & Sons, New York
11. Kuchanov SI (1973) Vysokomol Soedin A 15:2140
12. Korolev SV, Kuchanov SI (1986) Vysokomol Soedin A 28:1006
13. Vasnev VA, Vinogradova SV, Markova GD, Voytekunas VYu (1997) Vysokomol Soedin A 39:412
14. Panyukov SV, Kuchanov SI (1992) Journ Phys (France) II, 2:1973
15. Gordon M (1962) Proc Royal Soc (London) A 268:240
16. Harris TE (1963) The theory of branching processes. Springer, Berlin Heidelberg New York
17. Kuchanov SI, Korolev SV, Panyukov SV (1988) Adv Chem Phys 72:115

18. Chompff AJ (1971) Glass points of polymer networks. In: Chompff AJ, Newman S (eds) Polymer networks. Plenum Press, New York, p 145
19. Bank L, Ellis B (1982) Polymer 23:1466
20. Flory PJ (1976) Proc Royal Soc (London) A 351:351
21. Dusek K (1987) Macromol Symp 7:37
22. Flory PJ (1953) Principles of polymer chemistry. Cornell Univ Press, Ithaca, New York
23. Merz E, Alfrey T, Goldfinger G (1946) J Polym Sci 1:75
24. Kuchanov SI, Korolev SV, Zubov VP, Kabanov VA (1984) Polymer 25:100
25. Kuchanov SI, Kok C, ten Brinke G, to be published
26. Kuchanov SI, to be published
27. Kuchanov SI (1992) Adv Polym Sci 103:1
28. Kuchanov SI (1996) Kinetics and statistics of reactions on macromolecules. In: Kuchanov SI (ed) Mathematical methods in contemporary chemistry. Gordon & Breach, Amsterdam, chap 5
29. Kuchanov SI (1992) In: Allen G (ed) Comprehensive polymer science, 1st suppl. Pergamon Press, New York, chap 2
30. Greszta D, Madrare D, Matyjaszewski (1994) Macromolecules 27:638
31. Malmstrom EE, Hawker CJ (1998) Macromol Chem Phys 199:923
32. Kuchanov SI (1993) Polymer Science 35:215
33. Mayo FR, Lewis FM (1944) J Am Chem Soc 66:1594
34. Frensdorff HK, Pariser R (1963) J Chem Phys 39:2303
35. Greenley ZR (1980) J Macromol Sci A 14:445
36. Slocombe RJ (1957) J Polym Sci 26:9
37. Kuchanov SI (1989) Macromol Symp 26:371
38. Kuchanov SI, Orlova ZV, Koscheleva AV, Gorelov YuP (1989) Vysokomol Soedin A 31:474
39. Price FP (1962) J Chem Phys 36:309
40. Seiner JA, Litt M (1971) Macromolecules 4:308
41. Korolev SV, Kuchanov SI (1982) Vysokomol Soedin A 24:647
42. Plochocka K, Harwood HJ (1978) ACS Polymer Preprints 19:240
43. Harwood HJ (1987) Macromol Symp 10/11:331
44. Semchikov YuD, Smirnova LA, Knyazeva TE, Sherstyanykh VI (1990) Eur Polym J 26:883
45. Semchikov YuD, Slavnitskaya NN, Smirnova LA, Sherstyanykh VI, Sveshnikov TG, Borina TI (1990) Eur Polym J 26:889
46. Smirnova LA (1991) PhD thesis, Moscow State University
47. Semchikov YuD (1996) Macromol Symp 111:317
48. Kuchanov SI, Russo S (1997) Macromolecules 30:4511
49. Tolle J, Wittmer P, Gerrens H (1968) Makromol Chem 113:23
50. Wittmer P, Schmitt BJ (1987) Makromol Chem 188:2467
51. Fayt R, Jerome R, Teyssie Ph (1982) J Polym Sci Polym Phys Ed 20:2209
52. Hashimoto T, Tsukahara Y, Tachi K, Kawai H (1983) Macromolecules 16:648
53. Barucha-Reid AT (1960) Elements of theory of Markov processes and their applications. McGraw-Hill, New York
54. Vasnev VA, Kuchanov SI (1986) In: Korshak VV (ed) Advances in polymer chemistry. Mir Publ, Moscow, chap 4
55. Kuchanov SI (1976) Dokl Acad Nauk SSSR 229:135
56. Kuchanov SI, Brun EB (1979) Vysokomol Soedin A 21:700
57. Flory PJ (1953) Principles of polymer chemistry. Cornell Univ Press, New York
58. Korolev SV, Brun EB, Kuchanov SI (1982) In: Lifshitz IM, Molchanov AM (eds) Academy of Sciences of the USSR, Pushino, p 63
59. Kuchanov SI (1979) Dokl Acad Nauk SSSR 249:899
60. Brun EB, Kuchanov SI (1979) Vysokomol Soedin A 21:691
61. Kuchanov SI, Povolotskaya ES (1982) Vysokomol Soedin A 24:2179
62. Kuchanov SI (1973) Vysokomol Soedin A 15:2140

63. Korshak VV, Vinogradova SV, Kuchanov SI, Vasnev VA (1976) J Macromol Chem C 14:27
64. Kuchanov SI (1976) Vysokomol Soedin A 18:1878
65. Good IJ (1960) Proc Cambridge Phil Soc 56:367
66. Korolev SV, Kuchanov SI, Slin'ko MG (1983) Polymer J 15:785
67. Gordon M, Parker TG, Temple WB (1971) J Combinatorial Theory B 11:142
68. Kuchanov SI, Korolev SV, Slin'ko MG (1983) Polymer J 15:775
69. Kuchanov SI (1987) Sov Phys Dokl 32:392
70. Korolev SV, Kuchanov SI, Slin'ko MG (1982) Dokl Akad Nauk USSR 263:633
71. Gordon M, Scantlebury GR (1964) Trans Faraday Soc 60:604
72. Gordon M, Parker T.G (1971) Proc Royal Soc (Edinburgh) A 69:181
73. Brun EB, Kuchanov SI (1979) Vysokomol Soedin A 21:1393
74. Dobson G, Gordon M (1964) J Chem Phys 41:2389
75. Korolev SV, Kuchanov SI, Slin'ko MG (1982) Dokl Akad Nauk USSR 262:1422
76. Kuchanov SI (1979) Dokl Akad Nauk USSR 249:899
77. Kuchanov SI, Povolotskaya ES (1982) Vysokomol Soedin A 24:2190
78. Kuchanov SI, Kholostiakov AG, to be published

Received: August 1999

Author Index Volumes 101–152

Author Index Volumes 1–100 see Volume 100

Subject Index

MIX
Papier aus verantwortungsvollen Quellen
Paper from responsible sources
FSC® C105338

If you have any concerns about our products,
you can contact us on
ProductSafety@springernature.com

In case Publisher is established outside the EU,
the EU authorized representative is:
Springer Nature Customer Service Center GmbH
Europaplatz 3, 69115 Heidelberg, Germany

Printed by Libri Plureos GmbH
in Hamburg, Germany